ISOENZYMES IN BIOLOGY AND MEDICINE

ISOENZYMES
IN BIOLOGY AND
MEDICINE

ALBERT L. LATNER

Professor of Clinical Biochemistry
University of Newcastle upon Tyne

ANDREW W. SKILLEN

Lecturer in Clinical Biochemistry
University of Newcastle upon Tyne

1968

ACADEMIC PRESS
London and New York

ACADEMIC PRESS INC. (LONDON) LTD
Berkeley Square House
Berkeley Square
London, W.1

U.S. Edition published by
ACADEMIC PRESS INC.
111 Fifth Avenue
New York, New York 10003

Library of Congress Catalog Card Number: 68–19257

PRINTED IN GREAT BRITAIN BY
W. S. COWELL LTD AT THE
BUTTER MARKET, IPSWICH

Preface

Since it became clear that many enzymes are made up of mixtures of closely related substances, the study of these so-called isoenzymes has become increasingly important in relation to both biology and medicine. We have attempted in this book to give an account of their study in relation to human beings, animals, insects, higher plants, fungi, protozoa, algae and bacteria. We have attempted to include as far as possible most of the available information about chemical structure, physiological aspects and metabolic role as well as the use of isoenzyme observations in genetic, ontogenic and phylogenic studies. We have also included a detailed account of their use in the diagnosis and pathogenesis of disease as well as in the control of therapy. As far as possible we have tried to describe the technical methods available in such a fashion that, given the appropriate equipment and technical ability, the reader should be able to make the necessary determinations himself.

Whilst there are a few other publications dealing with special aspects of the field of isoenzyme studies, we believe that the time is now ripe to write about them in relation to the whole field of biology including medicine.

Publications dealing with isoenzymes are appearing at a remarkably rapid rate. In the text of this monograph we have been able to incorporate some thousand references. In spite of this, another three hundred or so have accumulated since the original manuscript was completed. As mentioned in the first chapter we have therefore done our best to bring the reader up-to-date by including, as an Appendix, an additional list of references with titles. It will be seen from this that work is advancing most rapidly in relation to the oxidoreductases, genetic applications, insect and plant studies and applications to clinical medicine. We feel, however, that nothing has appeared recently which seriously alters the general accuracy of statements made in our manuscript.

This monograph is primarily intended for biochemists, biologists, clinical biochemists and physicians. Whilst it is aimed at the postgraduate worker, it is hoped that it will also serve as a useful reference book for the undergraduate student in any of these fields.

We wish to thank the staff of Academic Press for the help they have given us and the patient way in which they have watched the manuscript develop. We are much indebted to our secretaries (Miss C. M. Harrison and Miss S. McVitie) for help in its production.

In conclusion we should like to thank the Medical Research Council for grants in aid of the research, which has continually stimulated our interest in this field.

March, 1968   A. L. Latner
                                                   A. W. Skillen

Acknowledgments

We wish to thank the authors whose names appear in the legends and the following editors and publishers of the various journals involved who gave permission to reproduce illustrations: The Editor, *Annals New York Academy of Science* (Figs 1, 6, 12, 25 and Table 3); Esevier Publishing Co., Amsterdam (Figs 3, 18, 21, 55, 56 and Table 12); The Editor, *Journal of Experimental Zoology* (Figs 4, 10, 17, 32 and 51); The Editor, *Journal of Neurochemistry* (Fig. 5); The Editor, *Experimental Cell Research* and the International Society for Cell Biology (Fig. 7); The Editorial Board, *The Biochemical Journal* (Figs 8, 16, 57, and 58, Table 5); The Editor, *Science* and the American Association for the Advancement of Science (Figs 9, 14, 30, 34 and Table 4); The Editor, *Proceedings of the National Academy of Sciences, U.S.A.* (Fig. 11); The Editor, *Biochemische Zeitschrift* (Figs 13 and 15); The Editor, *Journal of Biological Chemistry* and the American Society of Biological Chemists, Inc. (Fig. 19, Tables 1, 6, 7, 8 and 10); The Editor, *Biochemical and Biophysical Research Communications* (Fig. 20 and Table 11); The Editor, *Nature* (Figs 22, 33, 35, 36, 37, 43, 44 and 45); The Editor, *Journal of Molecular Biology* (Fig. 23); The Editor, *Bacteriological Reviews* and the American Society for Microbiology (Figs 24, 26, 27 and 28); The Editor, *Comparative Biochemistry and Physiology* (Fig. 31); The Editor, *Genetics* (Figs 38, 39, 40 and 52); The Editor, *Progress in Medical Genetics* (Figs 41 and 42); The Editor, *Hereditas* (Fig. 46); The Editor, *Journal of Embryology and Experimental Morphology* (Figs 47 and 48); The Editor, *Developmental Biology* (Fig. 49); Federation of American Societies for Experimental Biology (Fig. 50); The Editor, Biochemistry and the American Chemical Society (Table 2); The Editor, *Angewandte Chemie* (Table 9).

Contents

CHAPTER I

Introduction

IT IS NOW well recognized that a large number of enzymes exist in multiple forms. This applies not only to tissues and tissue extracts, but also to enzyme proteins which have been isolated in the crystalline state and are really mixtures. Isoenzymes are examples of these multiple forms. Precise definition of the word "isoenzyme"* is, however, rather difficult. Different tissues of the same individual or even of different species may possess closely similar enzymes, which are not really iso-enzymes. For the time being, most authorities believe that a broad definition such as "different proteins with similar enzymatic activity" best suits the current state of our knowledge. It is customary, for the most part, to limit this definition to multiple enzymes obtained from one tissue of one individual animal or plant or possibly a small organ, or a culture of a unicellular organism. An exception would be the major multiple forms of human alkaline phosphatase. They are nevertheless referred to as isoenzymes.

The definition, where appropriate, could well be extended to include the fact that the different proteins should have the same coenzyme. Even this limitation does not necessarily avoid difficulty. It would lead, for example, to the recognition of the isoenzyme relationship of the NAD-dependent malate dehydrogenases of animal tissues. One of these enzymes occurs in mitochondria and the other in the soluble cytoplasm. It has been shown that the purified mitochondrial component can give a number of different forms on starch gel (Thorne *et al.*, 1963). These are obviously isoenzymes but their properties differ so greatly from the cytoplasmic component that it has been suggested that the latter cannot be regarded as having an isoenzymic relationship to them (Kaplan, 1963). On the other hand, the two forms of NADP-dependent malate dehydrogenase of the mouse have been regarded as isoenzymes (Henderson, 1966); here again one occurs in the soluble cytoplasm and the other in the mitochondria. A similar distribution occurs with the two forms of the NAD-dependent malate dehydrogenase in *Saccharomyces cerevisiae*. These have also been regarded as isoenzymes (Witt *et al.*, 1966). It is interesting to note that quite a number of enzymes have been stated to exist in isoenzymic forms in the supernatant and mito-

---

* The word "Isoenzyme" was first used by Wróblewski and Gregory (1961) and is preferable to "isozyme" (Markert and Møller, 1959).

chondrial fractions of tissues. When more extensive studies of their properties have been made, this opinion may well have to change.

Recognition of the isoenzyme nature of certain proteins becomes much easier if the overall molecular structure is known. This might, as with lactate dehydrogenase or bacterial alkaline phosphatase, be made up of different combinations of the same number of specific polypeptide subunits. It might vary, as with bovine carboxypeptidase A, by means of simple amino acid replacement (Walsh *et al.*, 1966). Recognition of isoenzymic nature may be simplified by genetic studies, which demonstrate that one or more of the enzymically active protein types occur in homozygotes but that heterozygotes contain mixtures of the homozygote types.

Proof of the existence of isoenzymes can be extremely difficult in the absence of any pointers to molecular structure, since the activities of a number of different enzymes may overlap, as for example with the esterases. The mere finding of multiple bands of activity after any kind of electrophoresis is therefore not sufficient proof. Moreover, multiple bands may be produced by combination with different non-enzymic serum proteins (Latner, 1966).

Our knowledge of isoenzymes is most detailed in relation to those of lactate dehydrogenase, but it is rapidly extending in relation to many others. The growth of the literature has been truly remarkable and a number of excellent reviews have already appeared (Wieland and Pfleiderer, 1962; Lawrence, 1964; Wilkinson, 1965a; Beckman, 1966; Shaw, 1965). We are beginning to know something of the molecular structure of these interesting substances. Many studies have been made in regard to their genetic control as well as to their ontogenesis. Species differences have been used to shed light on evolutionary pathways. Much is being learned of their metabolic significance, especially in relation to such phenomena as feed-back inhibition, substrate inhibition and shuttle-mechanisms between mitochondria and soluble cytoplasm.

Our knowledge has been extended, not only in relation to isoenzymes in animals, but also to those in plants, fungi, protozoa and bacteria. In fact, it will not be too long before the whole range of life-forms has been covered.

Studies in human beings have already yielded data suitable for application in a number of clinical situations. Lactate dehydrogenase isoenzymes in the blood have, for example, already taken a significant place in the diagnosis of myocardial infarction, and in certain liver diseases. The meticulous genetic studies of Professor H. Harris and his colleagues, in relation to the distribution in blood of certain isoenzymes and their variant forms, promise to produce isoenzyme "fingerprints" of individuals and might even shed some light on that somewhat nebulous clinical entity "constitution"*. A number of other workers have made use of isoenzymes in the study of cancer.

All these aspects have been considered in the various chapters of this monograph. Because of the rapid growth of knowledge in this field, the authors have asked the publishers to include, as an appendix, a list of publications which have appeared since the manuscript was submitted. This list includes titles and it is hoped it will be of some additional use to the reader.

* See Harris, H. (1966); Appendix p. 232.

CHAPTER II

Isoenzymes of the Oxidoreductases

1. LACTATE DEHYDROGENASE

NIELANDS (1952) was the first to show that a beef heart lactate dehydrogenase preparation could be separated into two protein components by Tiselius electrophoresis. Each of these components had enzyme activity. Using high voltage membrane foil electrophoresis, Wieland and Pfleiderer (1957) later described up to five or six components in extracts of a number of rat tissues. About the same time Sayre and Hill (1957) and Vesell and Bearn (1957) by means of continuous flow paper electrophoresis and starch block electrophoresis respectively were able to separate three lactate dehydrogenases from normal human serum. Making use of so-called enzymoelectrophoresis, Wieme (1958) was the first to demonstrate five lactate dehydrogenases in this fluid.

It is now known that lactate dehydrogenase exists as five isoenzymes of differing sub-unit composition (see p. 17). They are each designated LDH-1, LDH-2, LDH-3, LDH-4 or LDH-5 in accordance with their mobility during electrophoresis in the pH range 7–9. LDH-1 is the most negatively charged under these conditions and so moves most rapidly towards the anode. The other suffix numbers increase as mobility of the isoenzyme towards the anode decreases.

Pfleiderer and Jeckel (1957) showed that the enzymes from pig, beef and rat heart, and from rat and dog skeletal muscles differ not only in their electrophoretic mobility but also vary in their turnover numbers, their pH optima, their inhibition by pyruvate or sulphite, their temperature coefficients and their reactivity with p-mercuribenzoate.

Using membrane foil electrophoresis, the distributions of the various forms of lactate dehydrogenase in the heart, liver, kidney and skeletal muscles of a wide range of vertebrates have been described (Haupt and Giersberg, 1958) and about the same time Wieland et al. (1959a) reported on the quantitative distribution patterns of lactate dehydrogenase in a variety of human and animal tissues. The range of animals studied by these two groups of workers included the dog, cat, mouse, dormouse, bat, rat, hamster, hedgehog, sheep, rhesus ape, tortoise, porpoise, toads, frogs, reptiles and fishes, domestic fowl, pigeons and other birds.

Starch gel electrophoresis was subsequently used in the investigation of the distribution of lactate dehydrogenase isoenzymes in the tissues

4

from a variety of animals. These included the pig (Markert and Møller, 1959), the mouse (Markert and Møller, 1959; Allen, J. M., 1961; Markert and Ursprung, 1962; Unjehm *et al.*, 1966), the rabbit (Markert and Møller, 1959), the chicken (Lindsay, 1963), the frog (Nace *et al.*, 1961), the rat (Tsao, 1960; Fine *et al.*, 1963), the bat (Manwell and Kerst, 1966), the guinea pig (Flexner *et al.*, 1960); the sheep (Masters, 1963; 1964) and the human being (Wróblewski and Gregory, 1961; Latner and Skillen, 1962). A comparative analysis of the isoenzyme patterns of extracts of heart and kidney from the mouse, hedgehog, pig, pigeon, frog, trout, mussel and blow-fly has shown that many of the patterns were species-specific (Agnall and Kjellberg, 1965).

A. DISTRIBUTION IN HUMAN TISSUES

A good deal of work on human tissues has been carried out by Wróblewski and his school (Wróblewski and Gregory, 1961). After electrophoresis, the starch gel was separated into segments, each of which was eluted and the lactate dehydrogenase activity determined in each extract. The distribution patterns studied included those from extracts of thyroid, cardiac muscle, lymph node, adrenal, lung, pancreas, kidney, spleen, skeletal muscle and liver (Fig. 1).

Similar patterns have been obtained by using visual staining techniques in which phenazine methosulphate has been utilized to transfer protons from $NADH_2$ to a tetrazolium salt (MTT or Nitro BT) with the production of an insoluble purple formazan in the form of bands occupying the positions of the isoenzymes (Latner and Skillen, 1961a). An example of the results obtained with extracts of various human tissues is shown in Fig. 2a. It can be seen that the liver extract is characterized mainly by the presence of the slowest moving form, whereas the major constituents of heart extract are LDH-1 and LDH-2.

Distribution patterns of LDH isoenzymes in individual human organs have been described using both visual staining and ultraviolet demonstration techniques after various forms of electrophoresis. The patterns include those of skin (Wieme, 1958; Weber and Pfleiderer, 1961; Carr and Skillen, 1963), testis (Blanco and Zinkham, 1963) and spermatozoa (Goldberg, 1963), placenta (Hawkins and Whyley, 1966) (see Fig. 2b).

Enzymoelectrophoresis in agar gel (Wieme, 1959a; Wieme and Van Maercke, 1961) and ordinary starch block electrophoresis (Plummer *et al.*, 1963) have also been employed in studies of the distribution of human tissue lactate dehydrogenase isoenzymes. Patterns in human tissues have been described by Hess and Walter (1960; 1961) who used column chromatography on DEAE cellulose and by Richterich *et al.* (1963) using chromatography on DEAE Sephadex (Fig. 3).

Using adsorption-elution on DEAE Sephadex, Richterich *et al.* (1963) could not detect variations in the lactate dehydrogenase isoenzyme patterns in different regions of normal human kidney. Regional variations have, however, been found following electrophoresis or differential assay of human (Ringoir and Wieme, 1965), rabbit (Jensen and Thorling, 1965) and rat (Fine *et al.*, 1963; Smith and Kissane, 1965; Thiele and Mattenheimer, 1966) kidney lactate dehydrogenases.

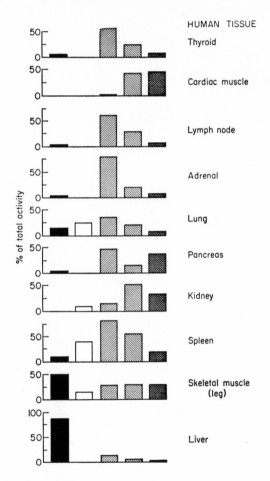

Fig. 1. Lactate dehydrogenase isoenzyme patterns of human tissues (reproduced with permission from Wróblewski and Gregory, 1961).

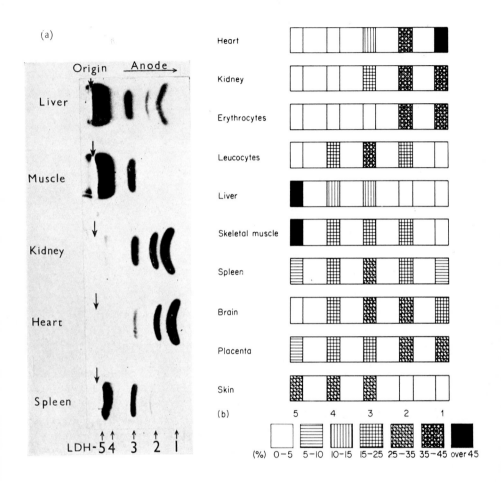

FIG. 2. Distribution of lactate dehydrogenase isoenzymes in human tissues. (a) Patterns visualized after starch gel electrophoresis. (b) Diagrammatic representation.

B

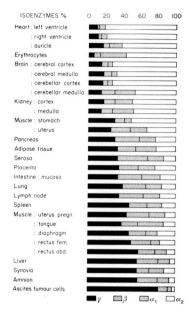

Fig. 3.   Human tissue lactate dehydrogenase isoenzyme patterns obtained by chromatography on DEAE Sephadex (reproduced with permission from Richterich *et al.*, 1963).

B. BLOOD LACTATE DEHYDROGENASE

The pattern of lactate dehydrogenase isoenzymes in human serum has been defined (Sayre and Hill, 1957; Vesell and Bearn, 1957; Wieme, 1959a; Latner and Skillen, 1961; Wróblewski and Gregory, 1961; Latner, 1962). Normal serum contains a relatively small amount of total activity, which tends to be highest in the positions corresponding to LDH-2 and LDH-3. It is possible, by using extraction methods after electrophoresis of sufficiently large amounts of serum, to demonstrate all five isoenzymes (Wróblewski and Gregory, 1961). Ultraviolet light detection methods or visual staining methods usually demonstrate bands corresponding to LDH-2 and LDH-3 with lesser amounts of LDH-1 and traces of LDH-5. Studies of the lactate dehydrogenase isoenzyme pattern of human cord blood have indicated a relative increase in the slower moving isoenzymes compared with adult blood serum (Wieme and Van Maercke, 1961).

The pattern of normal human serum is of particular importance in relation to the clinical application of isoenzyme patterns (Chapter IX). It is also important that any effect that serum may have upon isoenzyme patterns of tissue extracts be well defined. Diseased tissues liberate their isoenzymes into the circulation and migration of the latter during starch gel electrophoresis will almost certainly be affected by proteins, such as those contained in the blood stream. It has been shown that during agar and starch gel electrophoresis the mobility of LDH-5 towards the anode is decreased when serum is added to the extract and that LDH-1 tends to spread out along with the serum albumin (Latner, 1966).

Serum LDH isoenzyme patterns in relation to clinical diagnosis were used by Vuylsteek and Wieme (1958, cited in Wieme and Van Maercke, 1961) to differentiate between myocardial infarction and benign pericarditis. One year later an increase in serum LDH-5 was detected in the serum of patients with acute liver necrosis (Wieme and Demeulenaere, 1959, cited in Wieme and Van Maercke, 1961). Previously Vesell and Bearn (1957) had found an increase in the serum lactate dehydrogenase which migrated with the α_1-globulin during starch block electrophoresis in a case of myocardial infarction and an increase in the isoenzyme migrating into the α_2-globulin in a case of leukaemia. The diagnostic implications of LDH isoenzyme patterns are discussed more fully in Chapter IX.

The serum pattern has also been investigated in the monkey, the pig, the rabbit and the rat (Lawrence et al., 1960) and the patterns in insect blood have been studied (Laufer, 1960; 1961; 1963).

General increases in all five serum isoenzymes have been found after exercise and training of rats, which means that not only skeletal muscle LDH is released into the circulation (Garbus et al., 1964). It is also of interest that no significant alteration in the serum LDH isoenzyme pattern has been detected in rats subjected to hyperbaric oxygenation and Noble-Collip drum shock (Komatsu and Michaelis, 1966). Although the total serum enzyme level was elevated, all fractions were increased to the same degree. A non-specific increase in all four LDH isoenzymes of dog serum has been reported after experimental haemorrhagic shock (Vesell et al., 1959).

A number of reports have appeared which deal with LDH distribution in haemolysates of human red blood corpuscles. Here three major bands can be demonstrated, viz. LDH-1, LDH-2 and LDH-3, of which LDH-1 is the most prominent (Vesell, 1961). Occasionally LDH-4 may be present. Species with nucleated erythrocytes, e.g. the duck or chicken, contain LDH-5 (Vesell and Bearn, 1962). It has also been demonstrated in nucleated cells from the erythrocyte series of humans and guinea pigs (Vesell, 1964).

Young cells of the human erythrocyte series contain relatively more LDH-3 and LDH-4 than mature erythrocytes (Rosa and Schapira,

1965). An increase in erythrocyte LDH-5 has been reported as indicating active or hyperactive erythroid tissue (Starkweather *et al.*, 1965).

Chromatography on DEAE Sephadex combined with electrophoresis on cellulose acetate strips has helped in the characterization of lactate dehydrogenase from human erythrocytes (Dioguardi *et al.*, 1964) and granulocytes (Dioguardi *et al.*, 1963). Examination of the lactate dehydrogenase isoenzyme patterns of human platelets and bovine lens fibres has shown that they both contain a predominance of LDH-3 with only faint traces of LDH-5 (Vesell, 1965a). Since platelets and lens fibres, like mature human erythrocytes, lack a nucleus, these results strengthen the case for association between LDH-5 and the presence of the cell nucleus.

Studies of the LDH isoenzymes of buffy coat cells and erythrocytes of different species have shown that in all species the red and white blood cells have different patterns and that there are marked species differences in the white blood cell isoenzyme patterns (Walter and Selby, 1966). In human blood platelets LDH-2 and LDH-3 are the most prominent isoenzymes (Hule, 1966).

C. DISTRIBUTION IN OTHER ANIMAL TISSUES

Lactate dehydrogenase isoenzyme patterns have been examined in the tissues of a wide variety of animals. In addition to the early comparative studies (Haupt and Giersberg, 1958; Wieland *et al.*, 1959a) and those related to the individual species already mentioned, there are reports on the rabbit (Plagemann *et al.*, 1960a; 1960b) and the cynomologus monkey (Wieme and Van Maercke, 1961). Studies in relation to specific aspects of the isoenzymes rather than their distribution patterns will be discussed in the appropriate sections.

Multiple forms of lactate dehydrogenase have been shown to be present in the tissues of the speckled trout *Salvelinus fontonalis* and the lake trout *Salvelinus namaycush* (Goldberg, 1966). Nine bands of the enzyme have been detected after acrylamide gel electrophoresis of tissues of the speckled trout but there was no evidence of tissue specific patterns (Goldberg, 1965a). Further investigations have indicated that these nine isoenzymes occur only in hybrid species while in the parent (homozygous) species there are the usual five forms of the enzyme (Goldberg, 1966) which show species and tissue-specific distributions. Markert and Faulhaber (1965) have studied the patterns in muscles of thirty species of fish and they were able to classify them into four groups based on whether the fish muscle contained one, two, three or five major isoenzymes (Fig. 4). Five LDH isoenzymes have been found only in three genera, *Clupea* (herring), *Alosa* (shad) and *Merluccius* (whiting).

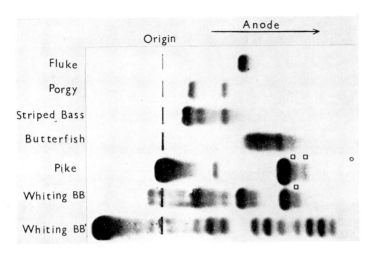

Fig. 4.   Lactate dehydrogenase isoenzyme patterns in muscle extracts of various species of fish. □, possible artefacts; ○, true minor isoenzyme (reproduced with permission from Markert and Faulhaber, 1965).

A single isoenzyme has been found in *Paralichthys dentalus* (hake) and two isoenzymes in many species of fish such as *Perca flavescens* (yellow perch), *Diplodus argentens* (bream) and *Mugil cephalus* (striped mullet). Genera such as *Cynoscion* (sea trout), *Roccus* (bass) and *Stizostedion* (pike) possess three isoenzymes. In many but not all fish there is evidence of some form of tissue specificity of the isoenzyme patterns. Whiting appears to be the most similar to mammals in its isoenzyme patterns (Markert and Faulhaber, 1965).

Three isoenzymes have been detected in the brains of a number of species of fish and Bonavita and Guarneri (1963a) have shown that the mobility of the predominant zone increases as one goes higher up the phylogenetic sequence. Vesell and Bearn (1962) have found six zones of lactate dehydrogenase in the erythrocytes of the carp.

Comparative studies of the lactate dehydrogenases in amphibia have shown that the kinetic properties of the enzymes in homogenates of the heart muscles are much more variable between species than the corresponding enzymes from skeletal muscles (Salthe, 1965). The lactate dehydrogenase isoenzyme patterns of *Rana pipiens* have been shown to be markedly different from those of *Rana catesbeiana* (Manwell, 1966). With the bullfrog, *R. catesbeiana*, the overall patterns of adult and tadpole heart and muscle are similar, with the adult having much greater total enzyme activities. Different patterns for heart and muscle LDH isoenzymes have been detected for both species of frog.

A study of lactate dehydrogenase isoenzyme patterns in birds has shown a correlation between the pattern and the flight habits of the individual species (Wilson *et al.*, 1963). The relative reactivity of extracts of the breast muscle of the birds with coenzyme analogues was used to identify the isoenzyme pattern in the muscle. An investigation of pea-fowl tissues has indicated that isoenzymes should be identified on the basis of properties other than electrophoretic mobility alone (Rose and Wilson, 1966).

D. NERVOUS SYSTEM

The heterogeneity of lactate dehydrogenase has been studied in the grey and white matter of human and sheep brains with a view to the possible clinical use of isoenzyme patterns in cerebrospinal fluid (Lowenthal *et al.*, 1961). More detailed studies on the isoenzyme distribution and the properties of the isoenzymes of nervous tissue have been reported (Bonavita and Guarneri, 1962; 1963a; 1963b). By means of agar-gel electrophoresis and elution of serial segments, kinetic studies have been made on the lactate dehydrogenases from the brains of various vertebrates (Bonavita and Guarneri, 1963a). A consistent phylogenetic sequence has been found through a series of eleven arbitrarily selected species (Fig. 5). Studies of the regional distribution in ox brain

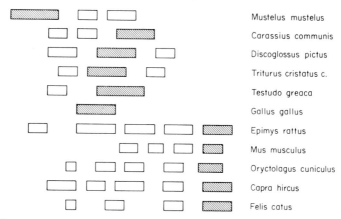

Mustelus mustelus
Carassius communis
Discoglossus pictus
Triturus cristatus c.
Testudo greaca
Gallus gallus
Epimys rattus
Mus musculus
Oryctolagus cuniculus
Capra hircus
Felis catus

FIG. 5.   Lactate dehydrogenase isoenzyme patterns in extracts of vertebrate brains (reproduced with permission from Bonavita and Guarneri, 1963a).

have shown three main types of pattern corresponding to the brain stem, the hemispheres and the striate body (Bonavita and Guarneri, 1963b).

Isoenzyme patterns of lactate dehydrogenase were at first believed to be the same in different regions of the human brain (Van der Helm, 1962a), although later studies using micro techniques (Van der Helm *et al.*, 1963) have shown slightly different distributions. The LDH iso-enzyme distribution has, however, been shown definitely to vary in extracts of different regions of the human brain when examined by agar gel electrophoresis (Gerhardt and Petri, 1965). If the relative proportions of the five isoenzyme fractions were used to calculate the relative amounts of the H and M types of LDH sub-unit present in different regions of the brain, it was possible to show a correlation between the H/M ratio and the oxygen supply to any particular region. Most regions of the brain did show a fairly uniform distribution with highest activities in LDH-1, LDH-2 and LDH-3. The dentate nucleus of the cerebellum was distinctive in having a very high H/M ratio. The lowest H/M ratios were found in the mammillary bodies, the optic chiasma, the lateral geniculate body and the lumbar spinal ganglion.

Investigations of the lactate dehydrogenase isoenzymes in cerebro-spinal fluid, blood, leucocytes and brain extracts (Van der Helm *et al.*, 1963) have shown that the isoenzyme pattern is virtually unchanged when the lactate dehydrogenase passes the blood/brain barrier.

More recent observations on the patterns of nervous tissue have shown that peripheral nerve and muscle show some similarities (Lowenthal *et al.*, 1964).

Differences in the distribution of the lactate dehydrogenase isoen-zymes between cutaneous and muscular nerves have been described in both cats and guinea pigs (Brody, 1966); nerves supplying the skin have been found to have less of the faster moving isoenzymes.

Using zone electrophoresis in starch paste, Futterman and Kinoshita (1959) have been able to separate five fractions of lactate dehydrogenase from rabbit retina. Three fractions have been found after ion-exchange chromatography of epithelial-endothelial extracts of rabbit cornea (Moore and Wortman, 1959). Electrophoresis on cellulose acetate has also been used in studies of patterns in the rat retina (Graymore, 1964; 1965). LDH-5 has been shown to be present in greater concentration in the retina of the young rat than in that of the more mature animal. A reduced level has been found in the retinas of young and adult "retinitis pigmentosa" rats. Graymore (1965) has suggested that the higher LDH-5 content of the developing rat retina is favourable for anaerobic metabolism and tissue differentiation, whereas in the "reti-nitis pigmentosa" animals the deficiency of LDH-5 does not favour anaerobic metabolism and normal differentiation is repressed.

Other workers (Bonavita *et al.*, 1963) have observed similar LDH isoenzyme patterns in the dystrophic rat retina. They have shown that at birth the retinas from normal rats and rats with inherited retinal degeneration cannot be differentiated by the LDH isoenzyme patterns but that significant differences between the two types are developed during maturation. The distribution of the two isoenzymes of malate dehydrogenase has not shown any differences in normal or dystrophic retinas (Bonavita, 1965).

E. TESTIS AND SEMINAL FLUID

In a variety of animals, including man, more than five forms of lactate dehydrogenase in mature testis and sperm can be demonstrated by gel electrophoresis (Blanco and Zinkham, 1963; Goldberg, 1963; 1964; Zinkham et al., 1964a). The isoenzyme patterns found in sperm provide evidence for a single cell type possessing multiple forms of enzymes. The additional lactate dehydrogenase isoenzyme (band X) of sperm has the same metabolic function as the other five, and in mature human and rabbit sperm it is the most active form of the enzyme (Zinkham et al., 1963). This "band X" is found in postpubertal testis and is present in differentiating spermatogonia and mature spermatozoa. Relative reaction rates with coenzyme analogues and α-hydroxy-acids other than lactate have helped to differentiate "band X" from other lactate dehydrogenase isoenzymes and it has been suggested that spermatozoa possess "band X" for necessary metabolic activity during passage from the testis to the site of fertilization in the oviduct of the female (Zinkham et al., 1963). "Band X" or LDH_x has been found as a single zone in human, rabbit, mouse and dog testes but as two zones in guinea pig and rat testes and as three zones in bull testes (Fig. 6).

Clausen and Øvlisen (1965) have studied LDH_x in human semen, seminal plasma and spermatozoal extracts using both electrophoresis techniques and kinetic studies with NAD analogues for its characterization. Using electrophoresis on cellulose acetate and acrylamide gel, Wilkinson and Withycombe (1966) have examined LDH_x in human, mouse and dog testes and have shown that it has a much greater affinity for 2-oxobutyrate than the other isoenzymes but that it resembles LDH-1 in its behaviour towards oxalate and urea, as well as in its temperature sensitivity.

Starch gel electrophoresis has been used in studies of LDH isoenzyme patterns in pigeon testes (Zinkham et al., 1964b); three types of pattern have been recognized containing seven, eight and four isoenzymes. These patterns are due to this lactate dehydrogenase being made up of three sub-units arranged in groups of four (see Section G).

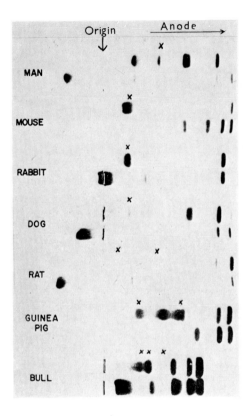

Fig. 6.   Lactate dehydrogenase isoenzyme patterns in extracts of mammalian testes. For each animal the pattern given by the testis enzyme is compared with that of another tissue from the same animal so that the mobilities of the normal isoenzymes and the testis-specific isoenzymes (×) can be compared (reproduced with permission from Zinkham *et al.*, 1964a).

F. TISSUE CULTURE STUDIES

Lactate dehydrogenase isoenzyme patterns have been examined in tissue culture cells (Philip and Vesell, 1962; Vesell *et al.*, 1962a; Nitowsky and Soderman, 1964). Typical patterns are illustrated in Fig. 7. It has been shown that chick embryo tissues in tissue culture reveal a common pattern with a marked preponderance of the two slowest moving isoenzymes (Philip and Vesell, 1962). Species specific patterns have been obtained in long term cultures of rabbit, chicken and human cells. They were independent of the organ of origin (Vesell *et al.*, 1962a). A number of clonal strains derived from liver have given patterns resembling that of fresh liver extract, but the patterns from cell lines derived from adult heart and embryonic intestine have not been characteristic of the tissue of origin. During serial propagation in vitro the changing patterns are characterized by the loss of the more rapidly migrating components (Nitowsky and Soderman, 1964). Examination of the lactate and other dehydrogenase isoenzyme patterns of several cultured cell lines, such as mouse lymphoblasts, hamster kidney fibroblasts, rat sarcoma cells and mouse skin fibroblasts has demonstrated the stability of isoenzyme patterns during long term culture; in general the slowest moving LDH isoenzyme has been the most prominent in all cell lines (Yasin and Goldenberg, 1966). The characterization of three human cell lines by biochemical parameters and chromosomal complement has shown that the lactate dehydrogenase isoenzyme patterns may be altered reversibly by changes in the mode of culture (German et al., 1964).

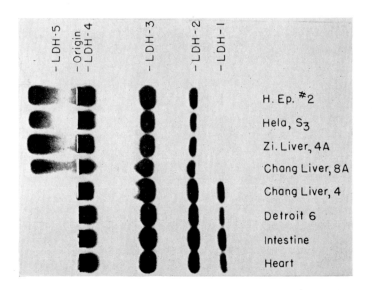

FIG. 7.   Lactate dehydrogenase isoenzyme patterns in cultured human cells (reproduced with permission from Nitowsky and Soderman, 1964).

Studies of changes in the patterns during culture of diploid and heteroploid cells from varied sources have indicated that there is a gradual disappearance of the faster moving isoenzymes as the cells age (Childs and Legator, 1965).

Elevation of the oxygen tension has enhanced the production of the faster moving isoenzymes in cultured heart cells and conversely low oxygen tension has increased the production of the slower moving isoenzymes (Goodfriend and Kaplan, 1963). All the changes reported have occurred within forty-eight hours and could be prevented by actinomycin. The rate of shift from the faster to slower moving isoenzymes in a tissue culture of chick heart cells could be retarded in a medium containing Krebs cycle acids or coenzyme A or by a high partial pressure of carbon dioxide (Cahn, 1963; 1964).

G. SUB-UNIT STRUCTURE OF LACTATE DEHYDROGENASE

Beef heart lactate dehydrogenase can be dissociated into four inactive sub-units by relatively high concentrations of guanidine or urea (Appella and Markert, 1961). Similar findings have been reported with respect to other mammalian lactate dehydrogenases (Cahn *et al.*, 1962; Jaenicke and Pfleiderer, 1962; Chilson *et al.*, 1964; Di Sabato and Kaplan, 1965; Withycombe *et al.*, 1965). Although urea denaturation was at first thought to be irreversible, Epstein and co-workers (1964) have been able to produce partially reversible denaturation of rabbit muscle lactate dehydrogenase using urea and β-mercaptoethanol and avoiding contact of the enzyme protein with glass.

When the curve obtained by plotting the reciprocal of the ratio of the lactate dehydrogenase activity in the presence and absence of urea, against the urea concentration is examined, a sharp inflexion can sometimes be seen, as shown in Fig. 8, which probably represents a sudden loosening of hydrogen bonds with subsequent unfolding of the enzyme molecule (Appella and Markert, 1961; Withycombe *et al.*, 1965).

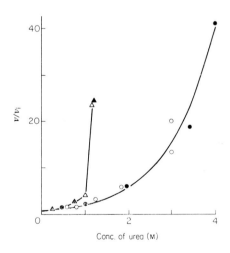

Fig. 8.   Effect of urea on LDH activity of human heart, $\bigcirc$; human liver, $\triangle$; ox-heart, $\bullet$; and rabbit muscle, $\blacktriangle$; preparations using 0·7 mM pyruvate as substrates. v/v_i, activity without urea/with urea (reproduced with permission from Withycombe *et al.*, 1965).

Failure to produce an enzymatically active protein after removal of urea has been attributed to the destruction of the helical structure of the sub-units (Appella and Markert, 1961; Withycombe et al., 1965).

The reduced and oxidized coenzymes ($NADH_2$ and NAD) and co-enzyme analogues have been shown to protect both chicken and beef heart lactate dehydrogenases from urea denaturation (Di Sabato and Kaplan, 1965). These findings were in agreement with those on sodium dodecyl sulphate denaturation of lactate dehydrogenase (Di Sabato and Kaplan, 1964), where the reduced coenzyme was most effective in protecting the enzyme from denaturation. The presence of a number of inorganic and organic salts is also effective, although the concentrations of the salts required is well in excess of the concentrations of nucleotides which are necessary for protection (Di Sabato and Kaplan, 1965).

The fact that there are five isoenzymes of beef lactate dehydrogenase has been explained on the hypothesis that each is made up of a group of four sub-units (Markert, 1962). The sub-units are polypeptide in nature and there are only two of them. If these are designated A and B then the five isoenzymes can be written B_4 (LDH-1), A_1B_3 (LDH-2), A_2B_2 (LDH-3), A_3B_1 (LDH-4) and A_4 (LDH-5). In this way one obtains five groups of four units with each group corresponding to the isoenzyme in parenthesis. A similar hypothesis using the symbols M and H for LDH-5 and LDH-1 respectively has been advanced to explain the five isoenzymes present in chicken tissues (Cahn et al., 1962). Tissues such as heart muscle contain a preponderance of H sub-units and tissues such as skeletal muscle a preponderance of M sub-units. It is now known that within a given species the two polypeptides have closely similar molecular weights and it is highly likely, therefore, that the different behaviour of the isoenzymes of lactate dehydrogenase during electrophoresis is largely conditioned by surface charge. It is just possible that molecular shape may also play some part but this seems to be unlikely. The sub-unit hypothesis is now supported by a good deal of experimental evidence, the phenomenon of dissociation followed by recombination (Markert, 1963a) and immunochemical tests (Cahn et al., 1962; Lindsay, 1963; Markert, 1963b; Plagemann et al., 1960b).

Markert (1963a) made a striking discovery when he mixed pure LDH-1 and LDH-5 in molar sodium chloride, froze and thawed the mixture and subjected the resultant solution to starch gel electrophoresis. A mixture of the five isoenzymes could be detected, in the concentrations expected from random recombination of the two sub-units. This is illustrated in Fig. 9. Following this demonstration of in vitro dissociation and recombination of equal amounts of LDH-1 and

LDH-5, Vesell (1965b) has described the patterns obtained by dissociation and recombination of varied proportions of LDH-1 and LDH-5. Patterns similar to those of liver, skeletal muscle, erythrocytes and kidney could be produced by mixing appropriate amounts of the two isoenzymes.

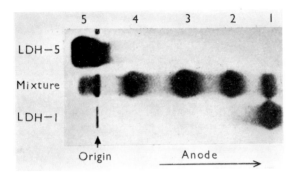

Fig. 9.   Dissociation and recombination of LDH-1 and LDH-5 from beef heart (reproduced with permission from Markert, 1963a).

Dissociation and recombination of lactate dehydrogenases has also been used to characterize the third type of LDH sub-unit present in testes LDH (Goldberg, 1965b; Zinkham *et al.*, 1963, 1964b). Dissociation and recombination of either the human or rabbit enzymes by freezing and thawing in 0·1 M phosphate buffer pH 7·0 containing 0·5 M sodium chloride has produced two new isoenzymes, one between LDH-2 and LDH-3 and the other between LDH-3 and LDH-4. No new isoenzymes could be detected when extracts of rabbit heart or skeletal muscle were subjected to this treatment. These results are taken to indicate that "band X" is made up of four similar sub-units designated "C" (Zinkham *et al.*, 1963). Studies of lactate dehydrogenase in pigeon testes have shown three types of pattern (Zinkham *et al.*, 1964b). In the first of these (Type I) there are seven isoenzymes, i.e. two "band X" isoenzymes. The second (Type II) showed eight isoenzymes and it has been suggested that it is most likely that there are five additional isoenzymes in this type but that two of these have the same mobility as the two slowest moving normal pigeon isoenzymes. In the third (Type III) only four isoenzymes could be detected. Dissociation and recombination of mixtures of Type I and Type III has yielded isoenzyme patterns identical with that of the Type II enzyme treated in the same manner. With Type I, dissociation and recombination produced a marked increase in the lesser of the two "band X" isoenzymes with a corresponding decrease in all the other non-"band X" isoenzymes. With Type III, similar treatment produced only very minor redistribution of enzyme activity throughout the four isoenzymes. Zinkham *et al.* (1964b) have suggested that these results could be explained if LDH synthesis in pigeon testes is controlled by three genetic loci, A, B and C. Although all pigeons are homozygous at the A and B loci, some are heterozygous at the C locus which controls band X. The five "band X" isoenzymes of Type II can be designated C_4, C_3C_1', C_2C_2', C_1C_3' and C_4'. The absence of other hybrids with the C and A or B sub-units is most probably due to the limited availability of these sub-units in mature testes (Zinkham *et al.*, 1966a). Investigations of approximately 1000 wild pigeons has revealed a mutation of the B locus giving a B polypeptide with electrophoretic mobility similar to that of the A polypeptide but with kinetic and heat stability characteristics similar to those of the B polypeptide (Zinkham *et al.*, 1966a). In tissues, other than testis, this results in three types of pattern designated Class I, Class II and Class III. Dissociation and recombination of mixtures of Classes I and III results in Class II (see Fig. 10).

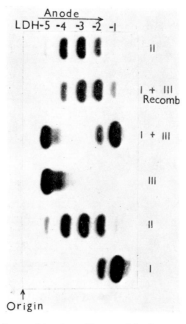

Fig. 10.   Dissociation and recombination of lactate dehydrogenases from pigeon heart. The Class II enzyme is produced *in vitro* by freezing and thawing a mixture of Classes I and III enzymes in 0·5 M NaCl/0·1 M phosphate (reproduced with permission from Zinkham *et al.*, 1966a).

(i) *Inter-species hybridization.* If lactate dehydrogenases from two species are mixed and frozen in molar sodium chloride solution, each enzyme dissociates into sub-units and when the mixture is thawed and recombination has occurred, hybrid enzymes can be detected.

The effect of pyridine nucleotides, temperature and different ions on *in vitro* hybridization of lactate dehydrogenases has been examined (Chilson *et al.*, 1965a). With mixtures of beef and chicken heart lactate dehydrogenases or those of chicken or beef heart and dogfish muscle, the enzymes would not hybridize (dissociate and recombine) without freezing the mixture of the two enzymes. This is the general rule. $NADH_2$ and coenzyme analogues have been shown to be effective in preventing denaturation of the enzyme during freezing and thawing. Only $AcPyNADH_2$ has been found to inhibit hybrid formation (Chilson *et al.*, 1965a).

Theories as to why hybridization occurs include the possibility that it may be a result of a combination of increased salt and protein concentration and/or decreased pH near the eutectic point. Chilson and co-workers (1965b) have demonstrated that it occurs most rapidly in the presence of both phosphate and a sodium halide. Freier and Bridges (1965) have indicated that magnesium and calcium ions are effective in its promotion, at a concentration one tenth of that of sodium chloride. Discrepancies between the observed and predicted isoenzyme patterns after dissociation and recombination may be due to varying rates of dissociation and reassociation between the different sub-units and different bond strengths between the sub-units in a system which has not reached equilibrium (Freier and Bridges, 1964). Arsenate and nitrate have been shown to promote hybridization of lactate dehydrogenase isoenzymes even in the absence of sodium chloride (Massaro and Markert, 1966).

Further studies on the reversible dissociation and hybridization of either lactate or malate dehydrogenases have indicated that they can be reversibly dissociated by guanidine hydrochloride, urea or an acid, and that this phenomenon is dependent on various ions and coenzymes or coenzyme analogues (Chilson *et al.*, 1966). The data obtained by these workers has shown that reactivation of the dissociated enzyme is time-dependent; no activity could be detected within one minute of diluting the urea treated enzyme with the reactivation solution and only 18% of the enzyme has been recovered in 150 minutes (Fig. 11). The best yields of reactivated MDH and LDH have been reported as being 75% and 40% respectively (Chilson *et al.*, 1966).

Reversible dissociation has been used to prepare interspecies hybrids of lactate, malate and triosephosphate dehydrogenases and as hybridization of dehydrogenases between widely divergent species, such as fish, amphibians, reptiles, birds and mammals, is possible, similar conformation of the enzyme molecules is probable (Markert, 1964; Chilson *et al.*, 1966).

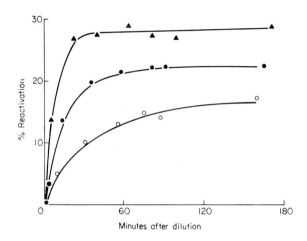

FIG. 11.   Effect of NAD and $NADH_2$ on reactivation of lactate dehydrogenase (H_4) from chick heart. Free enzyme was dissociated with 7·6 M guanidine in Tris-HCl buffer (pH 7·5). Reactivation was initiated by 50-fold dilution with 0·1 M Tris-HCl, (pH 7·5) plus 0·1 M β-mercaptoethanol, ○; or plus 1·5 mM NAD, ●; or plus 1·3 mM $NADH_2$, ▲ (reproduced with permission from Chilson *et al.*, 1965).

(ii) *Immunological studies.* Nisselbaum and Bodansky (1959) have examined the reactions of various rabbit lactate dehydrogenases with a rooster antiserum to rabbit muscle lactate dehydrogenase. Under certain specified conditions, the LDH activities of rabbit skeletal muscle and liver were completely inhibited, whereas those of rabbit spleen, kidney and serum were partially inhibited, and that of heart muscle only very slightly.

Antisera to human heart and liver dehydrogenases have been prepared both in roosters and rabbits. The antisera to human liver LDH have been shown to inhibit strongly the enzyme from both liver and skeletal muscle but have little effect on the enzymes from other organs. Similarly, the antiserum to human heart inhibited the enzymes from human heart, kidney, prostate, brain and erythrocytes but had little effect on those from liver or skeletal muscle (Nisselbaum and Bodansky, 1961).

Using rabbit antisera to the chicken M_4 and H_4 isoenzymes, Cahn *et al.* (1962) and Lindsay (1963) have not been able to detect any cross reaction. These workers have also described quantitative differences in the precipitation of the hybrid enzymes $H M_3$, $H_2 M_2$, and $H_3 M$, although both antibodies could completely precipitate each hybrid. Nisselbaum and Bodansky (1963) have reported similar findings with human LDH.

Antibodies to beef M_4 LDH have shown 86%, 68%, 41%, 23% and 0% inhibition respectively of the M_4, M_3H, $M_2 H_2$, $M_1 H_3$ and H_4 beef isoenzymes (Kaplan and White, 1963). Similar results, illustrated in Fig. 12 have been reported by Markert and Appella (1963). The combination of LDH and antibody has not been significantly affected by

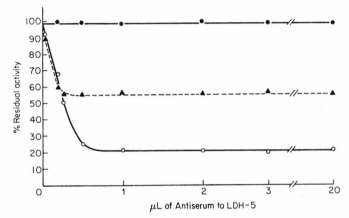

Fig. 12. Effect of mixing rabbit antiserum to beef muscle LDH-5 with preparations of LDH-1 (●); LDH-3 (▲) and LDH-5 (O) from beef heart and skeletal muscle (reproduced with permission from Markert and Apella, 1963).

the presence of substrate or coenzyme. No cross-reaction of LDH from beef, pig, mouse and chicken tissues could be detected (Markert and Appella, 1963) although such cross-reactions have been observed (Kaplan, 1963). Double-diffusion and immuno-electrophoresis techniques have enabled Avrameas and Rajewsky (1964) to demonstrate that anti-pig heart LDH and anti-beef heart LDH both react with pig heart and beef heart isoenzymes, as well as with LDH-1, LDH-2, LDH-3 and LDH-4 of human brain. Further studies (Rajewsky et al., 1964) have confirmed these findings and reaffirmed that the heart LDH of certain species cross-reacts and that the hybrid isoenzymes are immunologically related to both the non-hybrid isoenzymes.

Using a micro-complement-fixation technique, Wilson et al. (1964) have surveyed the evolution of lactate dehydrogenase from an immunological point of view. With rabbit antiserum to chicken H_4 lactate

c

dehydrogenase isoenzyme, they have been able to show close immuno-logical similarities between the H_4 isoenzymes from the chicken, turkey, duck, pigeon and ostrich. Those from reptiles, frogs and fish have been shown to be increasingly less cross-reactive with chicken H_4. These authors also described similar observations with chicken M_4 antiserum and the M_4 isoenzymes from these species. Studies with antisera to the fish isoenzymes have shown immunological changes during evolution from jawless to bony fishes (Wilson et al., 1964).

All these immunological studies have shown that, although heart muscle and lactate dehydrogenases from a variety of species do show close similarities, it is possible to detect small but significant differences in the extent of their immunological cross-reaction. These differences may be used in characterization of particular isoenzyme types.

(iii) *Analytical studies.* The first amino-acid analysis of a lactate dehydro-genase was reported by Gibson et al. (1953) who analysed a crystalline enzyme from rat liver. High voltage paper electrophoresis has been used in the analysis of the peptides produced by tryptic digestion of various lactate dehydrogenases (Wieland et al., 1960; Wieland and Pfleiderer, 1961). It has been found that the tryptic peptides from rat and pig heart are very similar but that those from rabbit and rat skeletal muscles are dissimilar. Small but significant differences have also been detected in the tryptic peptides from the lactate dehydro-genases of rat skeletal muscle and rat liver.

Examination of the amino-acid composition of the four isoenzymes from rat heart has shown significant regular differences between each of them. Wachsmuth et al. (1964) claim to have purified a number of lactate dehydrogenase isoenzymes viz. LDH-1, LDH-2 and LDH-3 from human brain, LDH-1 and LDH-2 from human kidney, LDH-1 and LDH-2 from human heart, LDH-5 from human liver, LDH-1 from pig heart and LDH-5 from pig skeletal muscle. Amino-acid analysis of the isoenzymes from different tissues of a particular species has shown that those of similar electrophoretic mobility have identical amino-acid composition. A gradual increase in the arginine, glycine, tyrosine and phenyl alanine content has been found going through the series LDH-1 to LDH-5. At the same time, there has been a corresponding decrease in the aspartic acid, alanine, valine and methionine content (Wachs-muth et al., 1964). These results are shown in Fig. 13. Another significant finding has been the much higher histidine level in pig LDH-1 than in pig LDH-5 or any of the human isoenzymes.

Pesce et al. (1964) have prepared crystalline M_4 and H_4 isoenzymes from both chicken and beef tissues and analysed their amino-acid

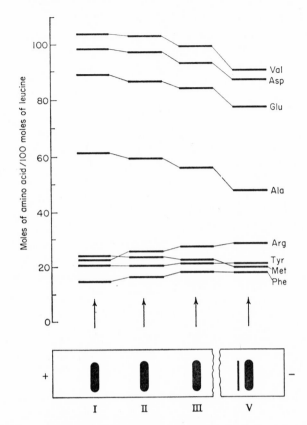

Fig. 13.   Amino-acid composition of lactate dehydrogenase isoenzymes from human tissues (reproduced with permission from Wachsmuth *et al.*, 1964).

composition (see Table 1). The average values for lysine, aspartic acid, glycine and valine have been shown to be fairly similar for all four. The methionine and isoleucine content of the chicken H_4 isoenzyme was lower than that of the other isoenzymes and the phenylalanine content of the beef H_4 was lower than that of the other isoenzymes (Pesce *et al.*, 1964). The chicken M_4 isoenzyme was outstanding in having a histidine content about three times that of the others with a corresponding decrease in the glutamic acid and tyrosine content. A comparative study of the M_4, H_4, HM_3 and $H_2 M_2$ isoenzymes from chicken tissues has shown that the amino-acid composition of H_4 from chicken heart and liver are practically identical (Table 2) and that the composition of $H_2 M_2$ and HM_3 agree well with the expected calculated values (Fondy *et al.*, 1964).

TABLE 1

Amino-acid composition of chicken and beef lactate dehydrogenases
(reproduced with permission from Pesce *et al.*, 1964)

Amino acid	Chick H	Beef H	Chick M	Beef M
Lysine	$99 \pm 4\cdot8$	$102 \pm 4\cdot8$	$112 \pm 8\cdot1$	$103 \pm 2\cdot0$
Histidine	$30 \pm 2\cdot5$	$26 \pm 2\cdot3$	$63 \pm 3\cdot5$	$33 \pm 1\cdot0$
Arginine	$35 \pm 1\cdot4$	$34 \pm 1\cdot5$	$35 \pm 2\cdot9$	$42 \pm 1\cdot3$
Aspartic acid	$129 \pm 4\cdot9$	$130 \pm 3\cdot2$	$125 \pm 4\cdot6$	$127 \pm 1\cdot9$
Threonine	$75 \pm 4\cdot4$	$57 \pm 1\cdot1$	51	$48 \pm 1\cdot3$
Serine	$107 \pm 8\cdot2$	$92 \pm 3\cdot7$	110	87
Glutamic acid	$122 \pm 2\cdot9$	$131 \pm 3\cdot1$	$102 \pm 3\cdot5$	$121 \pm 3\cdot5$
Proline	$38 \pm 4\cdot1$	$46 \pm 1\cdot6$	$44 \pm 3\cdot2$	$51 \pm 3\cdot3$
Glycine	$96 \pm 4\cdot2$	$94 \pm 1\cdot3$	$104 \pm 3\cdot7$	$100 \pm 2\cdot6$
Alanine	$88 \pm 0\cdot9$	$74 \pm 2\cdot2$	$81 \pm 3\cdot2$	$78 \pm 1\cdot3$
Valine	$125 \pm 3\cdot4$	$127 \pm 6\cdot4$	121 ± 10	$115 \pm 8\cdot3$
Methionine	$25 \pm 0\cdot7$	$34 \pm 1\cdot4$	$31 \pm 1\cdot4$	$32 \pm 1\cdot7$
Isoleucine	$66 \pm 7\cdot0$	$92 \pm 1\cdot5$	$85 \pm 5\cdot9$	$91 \pm 4\cdot3$
Leucine	$149 \pm 4\cdot9$	$139 \pm 2\cdot5$	$121 \pm 3\cdot0$	$136 \pm 2\cdot2$
Tyrosine	$31 \pm 1\cdot2$	$27 \pm 1\cdot4$	$19 \pm 0\cdot8$	$29 \pm 1\cdot4$
Phenylalanine	$19 \pm 0\cdot5$	$21 \pm 0\cdot7$	$27 \pm 1\cdot8$	$29 \pm 0\cdot8$
Cysteine	27	17	26	26

TABLE 2

Comparison of the amino-acid composition of the H_4 lactate dehydrogenases
from chick heart and liver (reproduced with permission from Fondy *et al.*, 1964)

	Heart	Liver
Lys	99	102
His	30	30
Arg	35	32
Asp	129	122
Thr	75	75
Ser	107	102
Glu	122	115
Pro	38	42
Gly	96	96
Ala	88	84
Val	125	135
Met	25	23
Ileu	66	73
Leu	149	146
Tyr	31	30
Phe	19	19

Examination of the fingerprint patterns obtained by tryptic digestion has indicated that the $H_2 M_2$ isoenzyme gives the same pattern as an equal mixture of H_4 and M_4 (Fondy et al., 1964). It has also been shown that H_4 and M_4 have many differences in amino-acid sequence. The tryptic digestion of each has yielded about forty peptides, thirty of which have been common to both H_4 and M_4 and perhaps ten characteristic for H_4 and ten others for M_4.

Evolutionary studies have shown that closely related species have similar amino-acid compositions for the H_4 as well as for the M_4 isoenzymes (Wilson et al., 1964; Kaplan, 1965). The histidine content of a number of different M_4 lactate dehydrogenases has been estimated and it has been shown that all birds have a high histidine content and amphibians a low histidine content (Wilson et al., 1964). The histidine content of the caiman, a reptile, is intermediate between that of birds and frogs, which is in agreement with accepted evolutionary concepts (Wilson et al., 1964; Kaplan, 1965). Other differences found have been that the arginine content of the frog enzyme is lower than that of birds and mammals and that the isoleucine content of mammalian LDH is higher than that of birds and frogs.

Di Sabato and colleagues (1963) have calculated the number of SH groups in crystalline preparations of LDH from various sources using data obtained by titration of the enzyme with p-chloromercuribenzoate in the presence of urea. Four SH groups are apparently involved with the active site of the beef H_4, chicken H_4 and chicken M_4 lactate dehydrogenases (Di Sabato and Kaplan, 1964). It has also been shown that these four groups are essential for binding the coenzyme and it is most likely there is one active SH group in each LDH sub-unit. Oxidized and reduced coenzymes, as well as certain coenzyme analogues, have been found to protect these SH groups against binding with p-hydroxymercuribenzoate (PHMB) (Di Sabato and Kaplan, 1965).

Further investigations on the function of sulphhydryl groups in lactate dehydrogenases and the amino-acid sequence around the essential group have been reported (Fondy et al., 1965). Nineteen different species of lactate dehydrogenase have been crystallized and treated with PHMB in 8 M urea. Although the usual number of thiol groups that bound PHMB varied from 16 to 26, only six binding sites could be detected in frog M_4 isoenzyme (Fondy et al., 1965). A ^{14}C-carboxymethylated peptide with electrophoretic and chromatographic properties identical with the active site peptide from frog muscle (M_4) lactate dehydrogenase has been found in tryptic digests of ^{14}C-carboxymethylated dehydrogenases from mammalian heart, bird muscle and heart and primitive fish M_4 (Fondy et al., 1965). In general, however, it appears that the overall primary amino-acid sequences in different lactate dehydrogenases have not been conserved during evolution. An identical peptide labelled with N-(N-acetyl-4-^{35}S-sulphamoylphenyl) maleimide has been isolated independently from pig heart LDH (Holbrook and Pfleiderer, 1965).

H. INTERACTION WITH COENZYME ANALOGUES

Anderson *et al.* (1959) have studied the effect of coenzyme analogues on the activity of the enzyme from rabbit muscle or bovine heart. They found that the rabbit muscle enzyme had a relatively greater affinity for coenzyme analogues than that from bovine heart. Further investigations have indicated that the relative reaction rates with the normal coenzyme and with various coenzyme analogues can be used to differentiate between lactate dehydrogenases of different origins (Kaplan *et al.*, 1960). For example, the ratio of the rate of reaction with 3-acetylpyridine adenine dinucleotide (APAD) to that with 3-thionicotinamide adenine dinucleotide (TNAD) has different values for each of the lactate dehydrogenases in extracts of human heart, liver, kidney and skeletal muscle (Kaplan *et al.*, 1960; Kaplan and Ciotti, 1961). This is shown in Table 3. The APAD/TNAD activity ratios for the lactate

TABLE 3

Lactate dehydrogenases of human tissues—comparison of the rate of reaction with 3-thionicotinamide adenine dinucleotide (TNAD) and 3-acetylpyridine adenine dinucleotide (APAD) (reproduced with permission from Kaplan and Ciotti, 1961)

| | Rate of reaction with APAD/ Rate of reaction with TNAD | | |
Heart	Liver	Kidney	Muscle
0·28	1·00	0·22	0·69
0·27	0·80	0·21	0·61
0·24	0·98	0·28	0·64
0·32	1·10	0·25	0·71

dehydrogenase of mammalian hearts are similar to each other and there is closer agreement between these values than there is between those for the heart and skeletal muscle enzymes from the same species (Kaplan *et al.*, 1960). Other observations have shown that the enzymes from fish heart muscle have a higher APAD/TNAD activity ratio than the enzymes from mammalian heart (Kaplan *et al.*, 1960). Studies of this activity ratio for the enzymes from the heart muscle and light and dark skeletal muscles of various species of fish have shown that in species such as the mackerel and the herring the ratios for heart and dark muscle are very similar and quite different from light muscle; whereas in other species such as the sea-robin and the salamander the ratios for dark and light muscle are similar to each other and very different from that for the corresponding heart muscle. Besides the APAD/TNAD activity ratio, other ratios, such as those between the activity with high and low concentrations of pyruvate or lactate and other analogues, such as pyridine-3-aldehyde nicotinamide-hypoxanthine dinucleotide have been used in comparative studies of the different lactate dehydrogenases from muscles of the phylum *Arthropoda* and from some annelids (Kaplan and Ciotti, 1961).

Coenzyme analogues have also been used to demonstrate differences between crude preparations of the enzyme from new-born and adult rat heart (Kaplan and Ciotti, 1961). Significant differences between the affinities of crystalline lactate dehydrogenase preparations from beef and chicken heart and skeletal muscle for various coenzyme analogues have also been found.

The ability of the enzyme to utilize coenzyme analogues has been used to compare the lactate dehydrogenases from the heart and skeletal muscles of a wide range of mammals, amphibia and birds (Cahn *et al.*, 1962).

A study of the relative reaction rates with the reduced form of nicotinamide-hypoxanthine dinucleotide and with $NADH_2$ has been used to show differences between the individual isoenzymes from bovine heart and skeletal muscle (Cahn *et al.*, 1962). These are shown in Table 4.

TABLE 4

Relative reaction rates of beef heart and skeletal muscle isoenzymes with reduced nicotinamide hypoxanthine dinucleotide ($NHXADH_2$) and $NADH_2$ (reproduced with permission from Cahn *et al.*, 1962)

| | $NHXADH_2$[a]/$NADH_2$[b] | |
	Heart	Muscle
LDH-1	2·50	2·78
LDH-2	1·74	1·83
LDH-3	1·29	1·23
LDH-4	0·78	0·70
LDH-5		0·53

[a] 3.3×10^{-4} M pyruvate as substrate.
[b] 10^{-2} M pyruvate as substrate.

Using chick lactate dehydrogenase isoenzyme preparations, Kaplan (1963) has compared the relative reaction rates of the five isoenzymes with the coenzyme and the coenzyme analogues, and shown that LDH-2, LDH-3 and LDH-4 give values close to those predicted from the values for LDH-1 and LDH-5.

I. HYDROXYBUTYRATE DEHYDROGENASE (HBD) ACTIVITY

The slower moving serum lactate dehydrogenase isoenzymes, when separated by continuous flow paper electrophoresis, have shown less activity with 2-oxobutyrate as substrate than with pyruvate as substrate, whereas the faster moving isoenzymes have the same activity with 2-oxobutyrate or pyruvate (Rosalki and Wilkinson, 1960). Further similar studies (Wilkinson et al., 1961; 1962; Plummer et al., 1963) have demonstrated relative activities with these substrates of extracts of various human, rabbit and mouse tissues (Table 5). The individual human LDH isoenzymes prepared by starch block electrophoresis have also been examined for their relative activity (Plummer et al., 1963a). In both

TABLE 5

Hydroxybutyrate and lactate dehydrogenase activities of various mammalian tissue extracts (reproduced with permission from Plummer et al., 1963)

| Species | Tissue | Dehydrogenase activity at 25°C (μmoles of $NADH_2$ oxidized/min/mg of N) | | Relative activities (b/a) |
		Pyruvate (a)	2-Oxobutyrate (b)	
Human	Heart	8·17	8·80	1·08
	Skeletal muscle	5·57	2·81	0·50
	Liver	5·23	1·69	0·32
	Serum (mean)	0·011	0·008	0·73
Rabbit	Heart	12·00	12·38	1·03
	Skeletal muscle	41·6	7·75	0·19
	Liver	5·07	1·73	0·34
	Kidney	2·64	2·35	0·89
	Serum	0·013	0·008	0·61
Mouse	Heart	0·445	0·155	0·35
	Skeletal muscle	0·230	0·033	0·14
	Liver	0·300	0·045	0·15
	Serum	0·041	0·011	0·27
Dog	Serum	0·053	0·012	0·23
Cat	Serum	0·048	0·014	0·29
Rat	Serum	0·199	0·038	0·19
Guinea pig	Serum	0·030	0·018	0·60

rabbit and human heart extracts the ratio of the activity with 2-oxo-
butyrate to that with pyruvate has been shown to be approximately
unity, whereas with the corresponding liver extracts, the ratio is approximately 0·17. With mouse heart and mouse liver extracts the ratio
has been found to be 0·35 and 0·15 respectively. Using sera from the
guinea pig, rat, cat, dog, mouse, rabbit and human, the ratio has been
shown to be 0·60, 0·19, 0·29, 0·23, 0·27, 0·61 and 0·73 respectively,
which has been said to suggest some species specific properties of lactate
dehydrogenase. With the preparations of individual isoenzymes the
ratio of 2-oxobutyrate to pyruvate activity is the same for the same isoenzyme from different tissues within a species and there is a gradual
regular increase in the ratio from 0·10 for LDH-5 to 1·0 for LDH-1.

Plummer and Wilkinson (1963) have estimated the temperature
coefficients, thermal stabilities and apparent enzyme-substrate dissociation constants of the lactate dehydrogenase from human tissues using
pyruvate, 2-oxobutyrate, lactate and 2-hydroxybutyrate as substrates.
The results have indicated that all the substrates are acted upon by the
same enzyme protein.

J. SUBSTRATE INHIBITION AND PRODUCT INHIBITION

The five lactate dehydrogenase isoenzymes show differing degrees of
substrate inhibition by pyruvate (Plagemann et al., 1960b). If the
enzyme assay is carried out at pH 7·0, LDH-5 is most active with
1·2 mM pyruvate; whereas LDH-1 is inhibited by this concentration
and shows only 70% of its optimal activity. If 0·15 mM pyruvate is used
as substrate, LDH-1 is most active, whereas this concentration is well
below the optimum for LDH-5. These properties can be used to determine the relative proportions of LDH-1 and LDH-5 in a particular
enzyme preparation. Similar observations have been reported by Cahn
et al. (1962) who have used the ratio of enzyme activity at 0·33 mM and
10 mM pyruvate respectively as an index of the proportions of the two
isoenzymes. The relationship between pyruvate inhibition and metabolic
role is discussed in Chapter V.

Similar inhibition by lactate is also well established, although it has
not been studied to the same extent as substrate inhibition by pyruvate.
Brody (1964) has used substrate inhibition by lactate as a means of
selectively visualizing isoenzymes after electrophoresis. If high (0·76 M)
lactate concentration is used in the incubation medium the faster-
moving isoenzymes are subject to substrate inhibition and do not show
up as well as the slower-moving entities.

Substrate inhibition of lactate dehydrogenase from human tissues has
been found to be pH dependent (Vesell, 1966). At pH 8·3, substrate

c*

inhibition of LDH-5 by lactate is negligible up to concentrations of the order of 0·3 M: whereas at pH 7 as much as 20% or 40% inhibition has been observed when 0·1 M or 0·25 M lactate respectively is used as substrate. Inhibition of LDH-1 by lactate does not show such marked pH dependence; within the pH range 6·6 to 8·3 there is marked substrate inhibition which increases from 15–25% at 0·1 M lactate to 70–80% at 0·25 M lactate. Similar observations have been made with LDH-5 from rat liver, LDH-1 from pig heart and LDH-5 from rabbit muscle (Vesell, 1966).

The well established differences in substrate inhibition of lactate dehydrogenase isoenzymes by pyruvate have received further attention following a report that this had no metabolic significance. This was based on some observations that the lactate dehydrogenase activities of whole human tissue homogenates behaved more similarly towards increasing concentration of pyruvate than would be expected from the differences in substrate inhibition of the individual isoenzymes LDH-1 and LDH-5 (Vesell, 1965c). This finding has not, however, been confirmed by other observers, who have shown (Fig. 14) that the lactate dehydrogenase activity of those human tissues containing predominantly the slower isoenzymes is inhibited significantly less than that of other tissues (Latner et al., 1966a). Further evidence that tissue homogenates

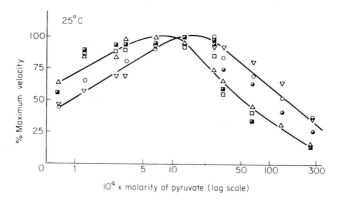

Fig. 14. Pyruvate inhibition of human lactate dehydrogenases (Latner et al., 1966a). ○, liver; △, heart; ▽, skeletal muscle; ◪, kidney; □, erythrocytes; ◑, spleen.

retain the substrate inhibition characteristics of their component isoenzymes has been presented (Stambaugh and Post, 1966b). It has also been shown that a change in temperature from 25°C to 37°C does not significantly affect the differences in substrate inhibition of purified LDH-1 and LDH-5 from rabbit muscle (Stambaugh and Post, 1966b).

This was also not in agreement with the finding of Vesell (1965c), who reported a smaller difference at 37°C.

Differences in product inhibition of the LDH isoenzymes from rabbit muscle have also been demonstrated (Stambaugh and Post, 1966a). Using 0·1 or 0·2 mM pyruvate as substrate, the activity of rabbit muscle LDH-1 at 25°C was inhibited some 60% by 20 mM lactate, whereas LDH-5 was inhibited by only 10% under the same conditions. With 10 mM lactate as substrate, the activity of LDH-1 at 25°C was inhibited 60% by 0·2 mM pyruvate, whereas LDH-5 was inhibited by some 40% under the same conditions. Further observations indicated that these differences were retained at 37°C even by whole tissue homogenates (Stambaugh and Post, 1966b).

K. KINETICS

Nisselbaum and Bodansky (1963) have prepared LDH-5 from human liver and a hepatoma, LDH-2 from human erythrocytes and LDH-1 from human heart and have examined their Michaelis constants for lactate, pyruvate and various coenzyme analogues (Table 6). Similar studies

TABLE 6

Michaelis constants of some human lactate dehydrogenase isoenzymes
(reproduced with permission from Nisselbaum and Bodansky, 1963)

	LDH-5		LDH-2	LDH-1
	Liver	Hepatoma	Erythrocytes	Heart
Nucleotides		moles $\times$ 10^4/l		
NAD^+	2·2	2·9	1·5	1·3
$TNAD^+$	1·2	1·6	0·96	0·80
3-$AcPyAD^+$	1·2	0·77	1·4	1·4
3-$PyAlAD^+$	5·4	5·3	10·4	5·9
3-$AcPyID^+$	2·2	1·8	2·4	3·3
3-$PyAlID^+$	11·1	9·3	13·4	13·5
NID^+	6·9	5·8	2·6	2·9
Substrates				
Pyruvate	3·1	3·4	0·59	0·68
L-Lactate at 4·8 $\times$ 10^{-4} M NAD^+	150	255	47	64
L-Lactate at 0·96 $\times$ 10^{-4} M NAD^+	155	190	84	55

(Nisselbaum et al., 1964) of the K_m(pyruvate), K_m(lactate), K_m(α-oxo-butyrate), K_m(α-ketovalerate), K_m(nicotinamide-inosine dinucleotide), K_s(pyruvate), K_i(oxalate) and K_i(oxamate) of LDH-1 and LDH-5 from human liver, brain and heart have confirmed that in a single species the isoenzymes have the same kinetic characteristics, regardless of their tissue of origin. The values of K_i(oxalate) and K_i(oxamate) for LDH-1 and LDH-5 have also been shown to be significantly different when either lactate or pyruvate were used as substrates (Table 7).

TABLE 7

Kinetic characteristics of some human lactate dehydrogenase isoenzymes
(reproduced with permission from Nisselbaum *et al.*, 1964)

	Substrate	LDH-5		LDH-1	
		Liver	Brain	Heart	Brain
K_m	Pyruvate	4·6	4·0	1·2	1·4
	α-oxobutyrate	63	63	17	16
	α-oxovalerate	101	89	54	51
	L-lactate	143	111	41	37
	L-α-hydroxybutyrate	48	52	44	64
K'_s	Pyruvate	168	156	58	46
K_i(oxalate)	Pyruvate	141	94	17·4	19·4
	Lactate	25	28	3·36	3·60
K_i(oxamate)	Pyruvate	142	94	23·6	31·0
	Lactate	380	422	202	140

K_m and K'_s, moles $\times$ 10^4/l.
K_i, moles $\times$ 10^{-5}/l.

Pesce *et al.* (1964) have prepared the crystalline beef and chicken isoenzymes and compared their optimal substrate concentrations K_m(pyruvate), K_m(lactate) and other catalytic characteristics (Table 8). Similar studies of the crystalline HM_3 hybrid from chick muscle, the H_2M_2 hybrid and H_4 isoenzymes from chick liver have been reported (Fondy *et al.*, 1964).

L. MISCELLANEOUS INHIBITORS

Wieland and Pfleiderer (1957) have shown that sulphite inhibits each of the five lactate dehydrogenase isoenzymes to a different extent. The faster-moving isoenzymes from rat tissues are more sensitive to this inhibitor (Wieland *et al.*, 1959b) LDH-1 being inhibited by as much as 80% under conditions where LDH-5 is inhibited by no more than 50%. Table 9 shows the relationship between electrophoretic mobility, sulphite inhibition and other physico-chemical properties of the isoenzymes.

TABLE 8

Kinetic characteristics of some crystalline beef and chicken lactate dehydrogenases (reproduced with permission from Pesce et al., 1964)

	Beef H_4	Chicken H_4	Beef M_4	Chicken M_4
Optimal pyruvate concentration	6×10^{-4} M	4×10^{-4} M	3×10^{-3} M	3×10^{-3} M
K_m pyruvate[a]	$1\cdot4 \times 10^{-4}$ M	$8\cdot9 \times 10^{-5}$ M	1×10^{-3} M	$3\cdot2 \times 10^{-3}$ M
Turnover number[b] with pyruvate at V_{max}	49,400	45,000	80,200	93,400
Optimal lactate concentration	4×10^{-2} M	3×10^{-2} M	2×10^{-1} M	$2\cdot5 \times 10^{-1}$ M
K_m lactate[a]	9×10^{-3} M	7×10^{-3} M	$2\cdot5 \times 10^{-2}$ M	4×10^{-2} M
$NHXDH_1:NADH_3$[c]	$2\cdot78$	$3\cdot02$	$0\cdot63$	$0\cdot40$
$AcPyAD_1:TNAD_1$[d]	$0\cdot17$	$0\cdot26$	$1\cdot0$	$4\cdot3$

[a] Determined by reciprocal plots.

[b] Represents moles of NADH oxidized per mole of enzyme per minute at 25°C at pH 7·5.

[c] Ratio of rates of reaction of NHXDH (reduced hypoxanthine analogue of NAD) at a pyruvate concentration of 3×10^{-4} M and of NADH at a pyruvate concentration of 1×10^{-2} M (1).

[d] Ratio of reaction rates at a lactate level of $1\cdot3 \times 10^{-2}$ M. AcPyAD, acetylpyridine analogue of NAD; TNAD, thionicotinamide adenine dinucleotide.

TABLE 9

Lactate dehydrogenases from rat tissues. Comparison of electrophoretic mobility, sulphite inhibition and other properties (reproduced with permission from Wieland et al., 1959b)

	LDH-5	LDH-4	LDH-3	LDH-2	LDH-1
Electrophoretic mobility at pH 8·6 (cm)	0	2	4	6	8
Sulphite inhibition (%)	31–37	42–48	47–53	62–78	69–75
Heat stability (half-life) denaturation at 50°C—mins)	3	10	40	100	∞
Optimal pyruvate concentration (mM)	1·2		⟶		0·15
Increase in rate of reaction with rise of 10°C	1·5		⟶		2·1

When pyruvate is used as substrate, oxalate behaves as a non-competitive inhibitor. The enzyme from human heart muscle is strongly inhibited by 20 mM oxalate, whereas that from human liver or skeletal muscle shows very little inhibition under these conditions (Emerson et al., 1964a). Examination of the effect of oxalate on the activities of purified isoenzymes from human tissues has shown a regular gradation from LDH-1 to LDH-5. A good agreement has been obtained between the observed values for oxalate inhibition and those calculated on the assumption that the inhibitor acts independently on the enzyme subunits. Oxalate is equally inhibiting with pyruvate and oxobutyrate as substrates.

Oxamate has been shown to be a competitive inhibitor of the reduction of 2-oxobutyrate and pyruvate. As oxobutyrate has been found to be the more weakly bound substrate, this inhibitor is more effective when 2-oxobutyrate is used as substrate (Plummer and Wilkinson, 1963).

Inhibition of human tissue lactate dehydrogenases by urea has also been investigated (Richterich et al., 1962; Richterich and Burger, 1963; Plummer et al., 1963; Withycombe et al., 1965). The electrophoretically slower moving isoenzymes, LDH-4 and LDH-5, are almost completely inhibited by 2·0 M urea, whereas the faster moving isoenzymes, LDH-1 and LDH-2, remain relatively unaffected by this concentration.

Using both pyruvate and 2-oxobutyrate as substrates and in some instances lactate and 2-hydroxybutyrate, Withycombe et al. (1965) have described an inverse relationship between sensitivity to urea inhibition and electrophoretic mobility of the LDH isoenzymes. With pyruvate as substrate, the lactate dehydrogenases of human liver, human heart, rabbit muscle and ox heart have been examined and the slowest moving isoenzymes found to be 50% inhibited by 1·0 M urea, whereas the fastest moving isoenzymes were 50% inhibited by 2·0 M urea. Using 2-oxobutyrate as substrate, the corresponding figures for 50% inhibition have been found to be 0·75 M and 1·25 M. The urea-analogues, methyl urea and hydantoic acid, have been shown to be even more effective inhibitors (Withycombe et al., 1965).

$NADH_2$ and pyruvate apparently have an effect on the urea-denaturation of lactate dehydrogenase isoenzymes (Lindy and Kontinnen, 1966a; 1966b). $NADH_2$ at a concentration of $1·3 \times 10^{-4}$ M reduces the urea inactivation of LDH-1 but does not have any effect on the inactivation of LDH-5. Increase in the pyruvate concentration also aids in the differentiation of LDH-1 and LDH-5 by urea inactivation. Incubation of human isoenzymes with 2·0 M urea for 10 minutes produces some 6% inhibition of LDH-1 and some 90% inhibition of LDH-5 when 10^{-3} M pyruvate is used as substrate. If, however, 5×10^{-3} M pyruvate is used, LDH-1 is activated by about 25%, whereas LDH-5 still shows 90% inhibition.

Two peptide inhibitors of lactate dehydrogenase have been isolated from human urine (Wacker and Schoenenberger, 1966). One of these is apparently specific for the muscle-type of lactate dehydrogenase and the other for the heart-type.

The specific activity of the LDH-5 isoenzyme of crystalline rabbit muscle enzyme can be increased by some 200% by means of sucrose density centrifugation or gel filtration in Sephadex G200 in the presence of low concentrations of β-mercaptoethanol (Gelderman and Peacock, 1965). This procedure separates a denser protein fraction which is inhibitory to LDH-5; the LDH-1 is not affected by β-mercaptoethanol under these conditions and is not inhibited by the denser protein fraction.

M. HEAT STABILITY AND OTHER PROPERTIES

The five rabbit lactate dehydrogenase isoenzymes have different heat stabilities (Plagemann et al., 1961). A previous report of varying heat stabilities of human serum LDH isoenzymes had been attributed to the protective action of certain serum protein fractions, such as albumin (Hill, 1958). Studies of the heat stabilities of LDH isoenzymes (Fig. 15) have led to the development of simple tests for determining the source of increased levels of serum lactate dehydrogenase (Wróblewski and Gregory, 1961).

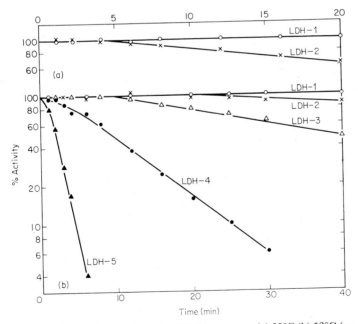

FIG. 15.   Effect of heat on the activity of dog LDH isoenzymes (a) 55°C (b) 53°C (reproduced with permission from Plagemann et al., 1961).

The effect of pre-incubation on the LDH and HBD activities of human heart or human liver extracts at temperatures from 30°–70°C has been examined (Plummer and Wilkinson, 1963). No significant variation could be detected up to 50°C but above this temperature the liver enzyme was shown to be considerably more labile (Fig. 16).

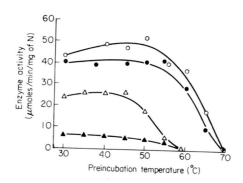

Fig. 16.   Effect of preincubation on the lactate dehydrogenase activity of human heart and liver extracts. ○, heart extract, pyruvate as substrate; ●, heart extract, 2-oxobutyrate as substrate; △, liver extract, pyruvate as substrate; ▲, liver extract, 2-oxobutyrate as substrate (reproduced with permission from Plummer and Wilkinson, 1963).

The optimal pyruvate concentration for both LDH-1 and LDH-5 has been shown to increase with increasing temperature (Plagemann et al., 1960b; Plummer and Wilkinson, 1963).

Calculation of the energies of activation of the five rabbit LDH isoenzymes has shown that the differences from $13,188 \pm 74$ calories for LDH-1 to 8285 ± 400 calories for LDH-5, show a linear correlation with electrophoretic mobility (Plagemann et al., 1960b). These results would seem to indicate that the turnover number for LDH-5 should be several thousand times greater than that for LDH-1. Plagemann (1960, quoted in Wróblewski and Gregory, 1961) has found evidence to suggest that the turnover numbers are, however, of the same order of magnitude.

N. MOLECULAR WEIGHT

The beef heart isoenzymes each have a molecular weight of about 134,000 (Markert and Appella, 1961). Dissociation using guanidine hydrochloride produces four sub-units of molecular weight $34,000 \pm 2000$ (Appella and Markert, 1961).

Ultracentrifugation studies of the pig heart isoenzyme have shown that this has a molecular weight of $115,000 \pm 6500$ (Jaenicke and

Pfleiderer, 1962). Similar studies and examination of the light-scattering and diffusion coefficients of the isoenzymes from human tissues have shown that these all have molecular weights in the range 117,800 ± 3800 (Jaenicke, 1963). Using similar techniques it has been shown that the limiting molecular weight of the sub-units of the heart-type isoenzymes from a variety of mammals was of the order of 36,000 (Jaenicke, 1964).

Chromatography on Sephadex G200 has shown that the molecular weight of mammalian isoenzymes is of the order of 110,000–120,000 (Wieland et al., 1963). Dissociation of the enzyme into sub-units with guanidine hydrochloride gave proteins with a molecular weight about half that of the original enzyme.

Pesce et al. (1964) have examined the physico-chemical characteristics of the beef and chicken H_4 and M_4 isoenzymes by a number of techniques (Table 10). The data indicate that the molecular weight of all four dehydrogenases is approximately 140,000 and similar values have been obtained for the crystalline enzymes from halibut or dogfish muscle and dog or human heart.

TABLE 10

Molecular weight and other physico-chemical properties of beef and chicken M_4 and H_4 isoenzymes (reproduced with permission from Pesce et al., 1964)

	Beef H	Beef M	Chick H	Chick M
Sedimentation coefficient $(s^0_{20,w} \times 10^{-13}$ cm s$^{-1})$	7·45	7·32	7·31	7·33
$s_{20,w}$ concentration dependence $(s_{20,w} \times 10^{-13}$ cm s^{-1} mg^{-1} ml$^{-1})$	0·064	0·033	0·043	0·034
s/D at the meniscus $\times 10^{-6}$	1·36	1·63	1·61	1·50
s/D (s from sedimentation velocity, D from maximum ordinate area method) $\times 10^{-6}$	1·28	1·25	1·31	1·36
s/D (s from sedimentation velocity, D from maximum ordinate method) $\times 10^{-6}$	1·36	1·35	1·40	1·46
Molecular weight from s/D	131,000	153,000	151,000	140,000
Molecular weight from s and D	123,000	116,000	123,000	128,000
Molecular weight from s and D	131,000	126,000	131,000	137,000
Diffusion coefficient $(D^0_{20,w} \times 10^{-7}$ cm^2 sec$^{-1})$	5·47	4·47	4·53	4·90
Frictional ratio, f/f_0, from $s^0_{20,w}$ and $D^0_{20,w}$	1·16	1·35	1·34	1·27
Partial specific volume, w_{20} ml g^{-1}	0·747	0·740	0·740	0·740

O. PHYSICAL AND CHEMICAL FACTORS AFFECTING ISOENZYME
PATTERNS

Although in the tissues of most vertebrates lactate dehydrogenase normally exists in the form of five isoenzymes, this pattern is subject to
genetic variation, which is discussed in Chapter VI. There are, however,
other factors chemical and physical in nature which can alter the patterns eventually obtained.

Vesell and Brody (1964) have demonstrated that the isoenzyme
pattern of human muscle stored for two years at $-25°C$ shows splitting
of the bands LDH-3, LDH-4 and LDH-5. Storage of muscle homogenates in the frozen state for several weeks has also produced similar
changes.

Fritz and Jacobson (1963) had previously reported that the slower
moving isoenzymes of mouse muscle could be split into sub-bands after
treatment with mercaptoethanol. They interpreted their results as indicating that these resulted from splitting off of coenzyme molecules
bound to the isoenzyme. Rabbit muscle lactate dehydrogenase showed
twelve fractions under the same conditions. Later observations (Fritz
and Jacobson, 1965) led to the rejection of the hypothesis that differential coenzyme binding was the correct explanation. No satisfactory
explanation is yet available but it appears that the SH groups of the
enzyme are involved.

Differences in lactate dehydrogenase isoenzyme patterns have been
observed when the tissues are extracted with different solutions. LDH-5
cannot be extracted as well with distilled water as with salt solutions so
that aqueous tissue extracts give a different isoenzyme pattern from
those made with physiological saline or buffer solutions (Vesell and
Brody, 1964). The relative insolubility of LDH-5 in water has also been
used by Fondy *et al.* (1965) in methods for the preparation of crystalline
isoenzymes.

The five isoenzymes of lactate dehydrogenase show different relative
positions after electrophoresis depending on the support medium and
the buffer solution used in the separation. If the sub-unit hypothesis (see
page 17) is correct, one would expect that the isoenzymes would be
equally spaced along the electrophoresis strip, since each differs from
the next in the series by the same increment of charge. Although this
type of pattern is often found after agar-gel electrophoresis (Wieme,
1959a; Kreutzer *et al.*, 1965), it is not always found after starch-gel
electrophoresis. Vesell (1962) has reported that the mobility of LDH-5
during the latter type of electrophoresis is dependent upon the concentration of that isoenzyme in the serum or tissue extract under investigation. In both starch- and agar-gels, the use of very dilute buffers

results in a tendency for LDH-5 to migrate towards the anode instead of the cathode (Ressler *et al.*, 1963a). It has been suggested that the electrophoresis media may contain anionic components which associate with the isoenzymes at low buffer concentrations. Patterns obtained in an agar plate which had been subjected to a preliminary electrophoresis have not been affected by buffer concentrations (Ressler *et al.*, 1963a). Washing of agar before electrophoresis (Kreutzer and Eggels, 1965) has also produced changes in the patterns with altered electroendosmotic flow. More recently a comparative study of the separations in Difco-Agar Noble, Behring Rein Agar, Oxoid Ionagar No. 2 and agarose has produced some surprising results (Kreutzer and Eggels, 1965). It appears that neither Behring Rein Agar nor Oxoid Ionagar are suitable for lactate dehydrogenase isoenzyme studies, since LDH-4 and LDH-5 are impeded except in the presence of an excess of another protein, such as γ-globulin. An association between the mobility of LDH-5 and γ-globulin had previously been described in dilute human tissue extracts (Carr and Skillen, 1963). Agarose is also unacceptable, since LDH-4 remains at the insertion point because of the slow endosmotic flow. More recently it has been shown that the addition of albumin to a final concentration $0 \cdot 1$ g/l to most commercially available forms of agar or agarose gives a suitable support medium for electrophoresis of LDH isoenzymes (Wieme, 1966).

The influence of differing concentrations of γ-globulin on the mobility of lactate dehydrogenase may have a marked effect on the detection of serum LDH-5 in patients with low or normal γ-globulin levels; only with relatively high serum γ-globulin levels can LDH-4 and LDH-5 be separated or move away from the starting slot when Behring Agar or Ionagar is employed (Kreutzer and Eggels, 1965). Difco-Agar Noble is apparently the best available for agar-gel electrophoresis as the five isoenzymes move equidistantly.

On the other hand, two-dimensional acrylamide gel electrophoresis (Raymond, 1964) has apparently indicated that the serum lactate dehydrogenase isoenzymes are not associated with any of the protein components of a single dimensional pattern.

Vesell and Bearn (1962) were the first to notice that the addition of coenzyme analogues increased the electrophoretic mobility of the LDH isoenzymes of haemolysates. Zondag (1963) made similar observations in relation to NAD and tissue homogenates when studying the effect of NAD on the cold stability of LDH-5. Kreutzer and Jacobs (1965) have reported that the addition of 10 mg NAD/ml serum produced changes in the electrophoretic mobility of the five isoenzymes. Five days after mixing serum and coenzyme, the mobilities of the five isoenzymes were significantly decreased; after 12 days the mobilities were approximately the same as those of untreated serum. Between the time of mixing and 12 days afterwards, the isoenzyme mobilities showed gradual alterations; the changes in mobility of LDH-2 seemed to be more marked than those of the other isoenzymes. If amounts of NAD varying over the range 2–10 mg/ml were added to the serum before electrophoresis, the shifts in mobility became more marked as the NAD concentration increased. Addition of $NADH_2$ produced similar but less marked changes.

Whilst examining a group of several thousand individuals, Kreutzer, Jacobs and Francke (1965) found five patients with serum LDH isoenzymes having abnormal mobilities during agar gel electrophoresis. In these five cases, LDH-4 moved more slowly than usual and LDH-5 more quickly. The other isoenzymes showed relatively normal mobility. If the abnormal sera were mixed with normal sera before electrophoresis, single zones of LDH-4 and LDH-5 with mobilities in between those of the normal and abnormal isoenzymes could be detected. The addition of NAD to the abnormal sera before electrophoresis caused the differences between the normal and abnormal mobilities to be less marked.

2. Malate dehydrogenase

Starch block electrophoresis of human serum has been used to separate three fractions of malate dehydrogenase (Vesell and Bearn, 1958). Starch gel electrophoresis has been used to fractionate the malate dehydrogenase activity in various crude animal tissue extracts (Markert and Møller, 1959). At least two bands could be detected. Similar results have been reported by Tsao (1960) and Latner and Skillen (1962). Two malate dehydrogenases have been demonstrated in human spermatozoa by acrylamide gel electrophoresis (Goldberg, 1963). Separations of malate dehydrogenases from rat and human tissues by cellulose acetate electrophoresis have indicated the presence of three isoenzymes, two of which are NAD specific and the third NADP specific (Sawaki et al., 1965). Using a similar electrophoresis procedure with a visual staining technique, Yakulis et al. (1962) have found up to four MDH isoenzymes in extracts of various human tissues, such as erythrocytes, kidney, liver, muscle, heart, lung and spleen; only the same two isoenzymes could be detected in most of these tissues.

Up to six fractions have been reported after agar gel electrophoresis of brain extracts and cerebrospinal fluid, three of these, however, had the same mobility as LDH isoenzymes and may not have represented true malate dehydrogenase (Lowenthal et al., 1961a; 1961b).

The heat stabilities of the various malate dehydrogenase fractions from human, rat and rabbit brains have also been investigated (Lowenthal *et al.*, 1961c). Human brain MDH was fairly heat stable; MDH1 and MDH6 were inactivated by heating to 55°C for 45 minutes, the other isoenzymes being unaffected by this treatment. The slowest moving MDH from rat and rabbit brain was the most heat-labile.

Malate dehydrogenase is distributed both in the mitochondrial and cytoplasmic cell fractions. Differences between the enzyme from these two sources have been described by a number of workers. Wieland *et al.* (1959b) showed that the mitochondrial and cytoplasmic enzymes could be separated by electrophoresis on membrane foil. The kinetic characteristics of the two enzymes have been shown to differ (Delbrück *et al.*, 1959a; 1959b). Grimm and Doherty (1961) have demonstrated that, after purification, the isoenzymes from ox heart could easily be separated by starch block electrophoresis in citrate buffer at pH 6·25. Thorne *et al.* (1963) have described multiple forms of pig heart mitochondrial malate dehydrogenase. As many as four bands have been detected after starch gel electrophoresis of the mitochondrial enzyme; the cytoplasmic enzyme occurred as a single zone under the same conditions.

Relative reaction rates of malate dehydrogenases with a number of coenzyme analogues have been used to differentiate between the soluble and mitochondrial enzymes of rabbit liver, brain, kidney, muscle and heart (Kaplan and Ciotti, 1961). Evolutionary studies have indicated that there is probably a characteristic species malate dehydrogenase which is modified in different tissues (Kaplan and Ciotti, 1961); in other words the malate dehydrogenases from the different tissues of a particular species appear more similar to each other than the enzymes from similar tissues in different species.

Relative reaction rates with coenzyme analogues have also been used to classify malate dehydrogenases in bacteria (Kaplan and Ciotti, 1961). Members of the *aerogenes-coli* group contain malate dehydrogenases which differ in kinetic properties from those in other bacteria. In general it would appear that the enzymes from closely related species are usually similar to each other, although some marked differences in those of the Bacillus group have been detected (Kaplan and Ciotti, 1961).

Studies of plant malate dehydrogenases have shown that in the higher plants they have similar kinetic characteristics with respect to their relative reaction rates with coenzyme analogues (Kaplan and Ciotti, 1961). Results with enzymes from yeast and *Neurospora* provide evidence that fungal malate dehydrogenases differ from those of green plants (Kaplan and Ciotti, 1961). These investigations, however, do not provide direct evidence of the presence of isoenzymes of malate dehydrogenase. It is only to be expected that the enzymes from different species would have properties peculiar to that species.

Recent observations of malate dehydrogenases from rat liver and rat kidney have shown that the monofluoro- and difluoro-derivatives of oxaloacetate may be used to discriminate between enzymes from different tissues (Kun and Volfin, 1966). The liver cytoplasmic enzyme reduces oxaloacetate and the two fluoro-derivatives with equal efficiency. It is not inhibited by monofluoro oxaloacetate concentrations of the order of 5×10^{-5} to 10^{-4} M. The enzymes from liver mitochondria and kidney mitochondria or cytoplasm with oxaloacetate or the difluoro-derivative as substrate are almost completely inhibited under these conditions. These results with fluoro-derivatives as substrates and inhibitors indicate that the mitochondrial malate dehydrogenase from rat kidney and liver resemble one another but that there are marked differences between the cytoplasmic enzymes of these tissues (Table 11).

TABLE 11

Inhibition of cytoplasmic and mitochondrial malate dehydrogenases
by fluoro-derivatives of oxaloacetate
(reproduced with permission from Kun and Volfin, 1966)

Preparation	V_{max} Ratios		
	OAA/F_1	OAA/F_2	F_2/F_1
Liver mitochondria	100	3·5	28
Kidney mitochondria	145	7·2	20
Liver cytoplasm	2·2	1·3	1·6
Kidney cytoplasm	14	2·7	5·0

3. ISOCITRATE DEHYDROGENASE

Isocitrate dehydrogenase from animal tissue has been separated into three components by starch gel electrophoresis (Markert and Møller, 1959; Tsao, 1960). In the rat, four different zones of activity have been found (Baron and Bell, 1962). Most tissues contain only one of these but heart and skeletal muscle have three, one of which is very weak. Three isoenzymes were detected in the human but no tissue examined had more than two of these. There were apparently four zones in human serum (Baron and Bell, 1962). Investigations using starch-gel electrophoresis at pH 6·2 rather than pH 8·6, at which the enzyme is unstable, showed only two bands of isocitrate dehydrogenase activity in human heart and liver, the faster component having half the mobility of albumin (Campbell and Moss, 1962). In liver extract, the faster band was predominant. In heart extract, the slower predominated.

The isoenzymes of isocitrate dehydrogenase at different intracellular localizations have been investigated. Immunological differences have been shown to exist between the intra and extra-mitochondrial components (Lowenstein and Smith, 1962). Two zones of enzyme activity in rat liver mitochondrial extracts have been demonstrated after starch-gel electrophoresis at pH 6·2 (Bell and Baron, 1964). Only the faster moving zones could be detected in the supernatant cell fraction. Isocitrate dehydrogenases in mouse tissues presented similar phenomena (Henderson, 1965). Mouse liver mitochondria contained two forms of the enzyme, one of which was identical with the supernatant enzyme; whereas mouse heart mitochondria contained only a single zone of enzyme activity.

Genetic variants of the supernatant enzyme (see Chapter VI) have been found in the livers of inbred strains of *mus musculus*. Three such variants have been demonstrated by electrophoresis and a sub-unit

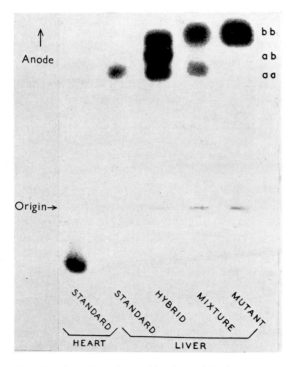

Fig. 17. Genetic variants of isocitrate dehydrogenase (reproduced with permission from Henderson, 1965).

structure for the enzyme has been proposed (Henderson, 1965). Two polypeptide sub-units sa and sb forming dimers giving the active iso-enzymes sa–sa, sb–sb, and sa–sb have been suggested. In the hetero-zygote, all three isoenzymes could be detected. In this case, that of intermediate mobility, sa–sb, was present in greatest amount (Fig. 17).

4. Galactose dehydrogenase

Starch gel electrophoresis has been used to demonstrate multiple forms of this enzyme in the livers of a variety of mammals (Cuatrecasus and Segal, 1966). Up to five components have been detected in extracts of rat liver, two in human and monkey liver and up to six in rabbit liver. If the extracts of monkey and human liver were mixed before electro-phoresis, the mobilities of the isoenzymes from each species did not alter and three isoenzymes could be demonstrated. This means that one isoenzyme of each species had a common mobility.

Ontogenic studies of the rat liver enzyme showed that the isoenzymes appeared in sequence, i.e. in the neonate only one isoenzyme could be detected, whereas in the 5- and 15-day-old rat the other isoenzymes appeared and became gradually increased.

5. Glutamate dehydrogenase

Using agar-gel electrophoresis and a visual staining method, it has been possible to separate the glutamate dehydrogenase of human tissues into five fractions; there appear to be differences in the isoenzyme patterns from different tissues (Van der Helm, 1962b).

6. Glucose-6-phosphate dehydrogenase

This enzyme was separated into multiple forms by starch-gel electro-phoresis (Tsao, 1960). A comparative study of mammalian erythrocytes has shown differences in the electrophoretic mobilities of the enzyme from the monkey, pig, rabbit and human being (Ramot and Bauminger, 1963). Investigations of the relationship between the electrophoretic mobility of the enzyme and erythrocyte glucose-6-phosphate dehydro-genase deficiency are discussed later (Chapter VI).

Removal of NADP from the erythrocyte enzyme by washing in 70% saturated $2 \cdot 7$ M ammonium sulphate containing $0 \cdot 27$ mM EDTA has produced a number of metastable subactive forms of the enzyme (Kirk-man and Hendrickson, 1962). Addition of NADP or warming to 25°C reactivates the enzyme. Sedimentation studies after removal of NADP have indicated a dimeric structure of the enzyme. Further studies have shown that the purified enzyme contains at least two polypeptide chains whose N-terminal amino acids are tyrosine and alanine (Chung and Langdon, 1963). The autosomally determined polymorphism of this enzyme in *Peromyscus* has provided evidence that the gene controlling glucose-6-phosphate dehydrogenase occurs as two alleles producing two polypeptide sub-units (Shaw and Barto, 1965). As a hybrid molecule is present in heterozygotes, a dimer structure for the enzyme is the most likely, e.g. the homozygotes can be designated as *aa* and *bb* and the heterozygote as *ab*.

Further evidence for the dimeric structure has resulted from the studies of Beutler and Collins (1965), who have produced a hybrid from rat and human erythrocytes by mixing the enzymes from each species during or after removal of NADP by repeated washing and dialysis, and then reactivating. The hybrid enzyme had electrophoretic mobility intermediate between the two pure forms.

7. PHOSPHOGLUCONATE DEHYDROGENASE

Visual demonstration of phosphogluconate dehydrogenases in red cell haemolysates, following electrophoresis in starch gels, has shown two bands of the enzyme which are genetically determined (see Chapter VI).

8. ALCOHOL DEHYDROGENASE

Starch gel electrophoresis of crystalline yeast alcohol dehydrogenase previously dialysed against phosphate buffer at pH 7·5 has revealed the presence of eighteen protein components, five of which had dehydrogenase activity (Watts and Donniger, 1962). If all metal ions were removed from the system, only two components could be detected. Using purified horse-liver alcohol dehydrogenase, four zones with enzymatic activity have been obtained (McKinley-McKee and Moss, 1965). The pattern was altered by the addition of oxidized or reduced coenzymes to the electrophoresis buffers. It has been suggested that metal ions and chelating agents alter the pattern by partial denaturation and inactivation of the enzyme.

Von Wartburg and co-workers (1965) have described an atypical human liver alcohol dehydrogenase with similar electrophoretic mobility to the normal enzyme but which differed in substrate specificity, sensitivity to inhibitors and pH optimum. Two fractions of alcohol dehydrogenase have been isolated from rhesus monkey liver using agar gel electrophoresis and chromatography on CM-cellulose (Papenburg *et al.*, 1965). The two fractions have been shown to differ in their kinetic properties and substrate phenomena.

9. Catalase

Although a number of reports of multiple forms of this enzyme in extracts of mammalian tissues have been described (Markert and Møller, 1959; Paul and Fottrell, 1961; Thorup *et al.*, 1961; Baumgarten, 1963), very little evidence of their isoenzymic nature was at first produced. Nishimura *et al.* (1964), however, have purified catalase from the liver and erythrocytes of both humans and rats and have shown the presence of at least three isoenzymes. Both human and rat liver catalase could be separated into two components by sucrose density gradient centrifugation. Using rabbit antiserum to human liver catalase, four components have been detected after immunoelectrophoresis of the human liver enzyme. Four components of the erythrocyte enzyme could also be detected with the corresponding antiserum. Rat erythrocyte catalase has been separated into three distinct pairs of arcs by similar techniques. The rat liver enzyme can be separated into two peaks by column chromatography on DEAE-celite-calcium phosphate gel. One of these shows three components and the other six when examined by immunoelectrophoresis. King and Gutmann (1964) have also described a rat liver catalase which gives three precipitin arcs after immuno-electrophoresis.

Thorup *et al.* (1964) have demonstrated that catalase from human erythrocytes could be separated into three peaks, A, B, and C, by chromatography on DEAE-cellulose. These three isoenzymes had different mobilities during starch-gel electrophoresis. Examination of their relative distributions in old and young normal erythrocytes, obtained by differential centrifugation, has indicated that the A fraction is increased and the C fraction decreased in younger erythrocytes. A similar isoenzyme pattern is found in patients with haemolytic disorders.

10. Multiple dehydrogenase activities

Koen and Shaw (1964) have presented evidence that several dehydrogenase activities are apparently associated with the same protein zones after starch gel electrophoresis. The five lactate dehydrogenase components could be easily identified and purple zones of formazan indicating enzyme activity could be detected in exactly the same positions as the LDH isoenzymes when glutamate, α-hydroxybutyrate, alanine and α-glycerophosphate were used as substrates. When α-glycerophosphate was used the major activity, however, was in a single zone well separated from the other isoenzymes and when the other three substrates were used two additional zones with the mobility of the two major malate dehydrogenase components could be detected.

Aspartate dehydrogenase could also be visualized as two zones with the same mobility as the malate dehydrogenase isoenzymes. Changes in the buffers used in the electrophoresis system did not alter the relative positions of the isoenzymes, so it would appear that these are true multiple dehydrogenase activities.

Among the dehydrogenases identified during such procedures was a component which migrated cathodically. This has since been identified as alcohol dehydrogenase; it produces reduction of tetrazolium salts, even in the absence of substrate and appears to be the major source of "nothing dehydrogenase" activity (Shaw and Koen, 1965).

Conklin *et al.* (1962) described seven fractions of LDH after starch gel electrophoresis with visual staining of the isoenzymes, two of which did not require NAD or phenazine methosulphate in the incubation medium. They have also indicated that after butanol treatment of tissue homogenates prior to electrophoresis only one isoenzyme could be detected and it was suggested that this indicated a possible association of the enzyme with lipids or lipoproteins. If the isoenzymes were due to binding of a single enzyme component to various lipoprotein fractions, multiple dehydrogenase activities could be easily explained by assuming that each lipoprotein band bound a number of different dehydrogenases. Ressler *et al.* (1963b) also examined LDH isoenzyme patterns in butanol extracted tissue homogenates and concluded that the results of Conklin *et al.* (1962) represented butanol inactivation. They could find no evidence of lipid or lipoprotein binding.

Katz and Kalow (1965) have examined the starch gel electrophoresis patterns of extracts of human tissues. Five bands of LDH, six of MDH, four of ICDH and one each of G6PDH and GDH could be detected. There was no evidence to support multiple dehydrogenase activities. Studies of the dehydrogenases in extracts of chick tissues have yielded similar results (Ecobichan and Kalow, 1966a). After starch gel electrophoresis of lactate dehydrogenase isoenzymes in rat tissues, Buta *et al.* (1966) were unable to show any activity of their eight LDH isoenzyme fractions with malate or α-glycerophosphate as substrate.

CHAPTER III

Isoenzymes of the Transferases

1. ASPARTATE AMINOTRANSFERASE

TWO FRACTIONS of aspartate aminotransferase have been detected after chromatography of soluble rabbit liver proteins on DEAE cellulose (Moore and Lee, 1960). Other workers have also demonstrated two aspartate aminotransferases in human, dog and pig hearts by the use of paper electrophoresis at pH 7·4 (Fleischer *et al.*, 1960). One moved as an anion and the other as a cation. The anionic fraction of the different species showed variable mobility. Using column chromatography on DEAE and CM celluloses, it has been shown that ox and pig heart contain two fractions of aspartate aminotransferase and there is evidence that one fraction is contained in the mitochondria and the other in the cytoplasm (Borst and Peeters, 1961). Similar findings in relation to the cattle and pig skeletal muscle enzymes have been reported (Kormendy *et. al.*, 1965).

Two isoenzymes from rat liver have been separated by means of agar gel electrophoresis at pH 7·5 (Boyd, 1961; 1962). The cathodic migrating component corresponded to the mitochondrial enzyme and the anodic one to the cytoplasmic enzyme. After experimental liver necrosis, the mitochondrial enzyme has been detected in rat serum, although it was not found in normal serum, which contained only the anodic component (Boyd, 1962).

Similar patterns in human material have been obtained after starch gel electrophoresis and detection by ultraviolet light or tetrazolium salt reduction (Boyde and Latner, 1962). Aqueous extracts of human liver, heart and kidney have shown the anodic and cathodic components. A second anodic component could be demonstrated when the tissue extracts were mixed with normal human serum. This has been shown to be due to binding of the mitochondrial enzyme by a serum protein (Latner, 1965). This second anodic component has been observed in serum from a patient with myocardial infarction and from a patient with carbon tetrachloride poisoning (Boyde and Latner, 1962). It has not been possible to demonstrate a fine structure in the rather broad bands obtained by the visual staining method (Boyde and Latner, 1962) although the use of a different staining medium containing a diazonium salt which couples directly with oxaloacetate has indicated that the major anodic component may be composed of up to three subfractions

(Decker and Rau, 1963). There appear to be species differences in the migration of the isoenzymes, since the soluble fraction enzymes of the human, rat, pig and dog had different mobilities.

Nisselbaum and Bodansky (1964) have prepared both the anionic and cationic components from human heart and liver, using chromatography on hydroxyapatite. They also immunized rabbits with preparations of each of the heart isoenzymes and studies with the antisera indicated that the anionic and cationic isoenzymes were immunochemically different, whereas the anionic isoenzyme from human heart could not be differentiated immunochemically from the corresponding isoenzyme in human liver.

Morino *et al.* (1963; 1964) have also found immunological differences between the mitochondrial and cytoplasmic components of bovine liver. They have made concentrated preparations of the enzymes from the mitochondrial and supernatant fractions of beef heart and liver, pig heart and liver, and rat liver, heart, kidney, brain and skeletal muscle. Antisera to the supernatant and mitochondrial enzymes from beef liver were prepared in rabbits. Using agar diffusion techniques, as well as tube-precipitation followed by enzyme assay of the supernatant, it has been possible to show that the enzymes of the soluble fractions were specifically inhibited by the antiserum to the component in the beef liver soluble fraction.

A considerable number of investigations on the kinetic characteristics of the mitochondrial and cytoplasmic enzymes have shown the existence of differences in the Michaelis constants for both L-aspartate and α-ketoglutarate. Both the K_m(α-ketoglutarate) and the K_m(L-aspartate) at fixed concentrations of L-aspartate and α-ketoglutarate respectively were significantly higher for the mitochondrial isoenzymes in both man and animal (Fleischer *et al.*, 1960; Borst and Peeters, 1961; Boyd, 1961; Nisselbaum and Bodansky, 1964). Similar Michaelis constants were found on the one hand for the mitochondrial isoenzymes and on the other for the cytoplasmic isoenzymes in all the species examined.

In contrast, Henson and Cleland (1964) have reported that the K_m(L-aspartate) values for the two aspartate aminotransferase isoenzymes from pig heart were not significantly different. The enzymes were prepared by ammonium sulphate fractionation, followed by chromatography on columns of CM cellulose. Other workers (Nisselbaum and Bodansky, 1966) have made similar investigations of the pig heart enzymes and have shown that one of the purification stages used by Henson and Cleland (1964) was deleterious to the mitochondrial enzyme. Moreover both peaks from the CM-cellulose column moved anionically during electrophoresis at pH 6·8–7·0. Pig heart mitochondrial enzyme prepared by other techniques moved cathodically during electrophoresis and its kinetic characteristics were of the same order as the mitochondrial enzymes from other species. These results have indicated that heating to 75°C destroys the mitochondrial component. This greater heat lability has also been described by Wada and Morino (1964).

Boyd (1966) has studied methods for the preparation of purified isoenzymes from rat tissues. Two components could be isolated from all tissues. Water and sucrose extractions were not too successful with the mitochondrial entities. Butanol extraction proved highly effective. Ammonium sulphate fractionation after butanol extraction of rat liver has been used to isolate isoenzyme preparations which were further purified by chromatography on DEAE-cellulose (Boyd, 1966).

Assay of an enzyme preparation at pH 6·0 and pH 7·5 could be used to estimate the relative proportions of the supernatant and mitochondrial enzymes (Boyd, 1966) since they had different pH optima.

Boyde (see Latner, 1965) has found that the human serum protein which binds mitochondrial aspartate aminotransferase has the same mobility as α_2-macroglobulin during electrophoresis and is probably identical with it (Fig. 18).

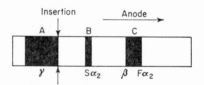

FIG. 18.   Aspartate aminotransferase isoenzymes; effect of mixing tissue extract with serum. A, mitochondrial isoenzyme; B, mitochondrial isoenzyme bound to serum protein; C, cytoplasmic isoenzyme.

2. HEXOKINASE

Using starch gel electrophoresis, Katzen et al. (1965) have separated multiple forms of this enzyme from human cell cultures and rat tissues. Two zones could be visualized after electrophoresis of extracts of the human cell culture material; four zones were present in the extracts of rat liver. K_m values of approximately 10^{-5} M, 10^{-4} M, 10^{-6} M and 10^{-2} M could be attributed to these four isoenzymes on the basis of visual staining in the presence of varying concentrations of glucose.

Gonzalez et al. (1964) have also demonstrated four hexokinases in rat tissue extracts by column chromatography on DEAE cellulose. Katzen and Schimke (1965) have described the patterns given by extracts of rat brain, kidney, muscle, liver and fat pad and have designated the four isoenzymes Type I to Type IV in order of increasing electrophoretic mobility. The Type I hexokinase was most predominant in brain and kidney and Type II in muscle and fat pad, whereas all four isoenzymes were present in liver (Fig. 19). These isoenzymes have been purified by a combination of ion-exchange chromatography, ammonium sulphate fractionation and gel filtration (Grossbard and Schimke, 1966). Kinetic studies have demonstrated that the isoenzymes, irrespective of

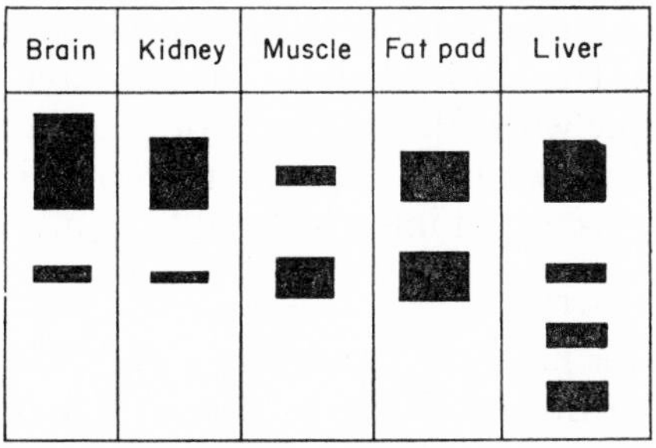

Brain	Kidney	Muscle	Fat pad	Liver

Fig. 19. Distribution of hexokinases in rat tissues (reproduced with permission from Grossbard and Schimke, 1966).

tissue of origin, differed in relation to their K_m values for glucose and ATP, their K_i values for glucose-6-phosphate and ADP, and their heat stabilities. However, they behave similarly with respect to pH optima and substrate specificity. They also had similar molecular weights.

Other mammalian species have also been shown to possess multiple forms of hexokinase (Grossbard et al., 1966). Very little difference between the tissue isoenzyme patterns from a variety of species has been detected and there does not seem to be any likelihood of species specificity in this respect.

3. CREATINE KINASE

Creatine kinase has been separated into multiple forms by agar-gel electrophoresis coupled with ultraviolet detection methods. Creatine kinase could be split into three entities by electrophoresis at pH 9·0. Two moved towards the anode and the third was cationic. The faster migrating anode component was found in brain and other central nervous system tissues and the cathode component in skeletal and heart muscle. All three components appeared in smooth muscle (Burger et al., 1964). Deul and Van Breeman (1964) confirmed the differences in mobility between human skeletal muscle and brain creatine kinase but found five zones in human cardiac muscle. Although the mobilities of the isoenzymes differed from species to species, their distribution in organs of members of the same species was always the same (Burger et al., 1964).

A considerable amount of creatine kinase activity has been located in the mitochondria of muscle and brain (Jacobs *et al.*, 1964). Although cytoplasmic creatine kinase isoenzymes showed variations in the distribution patterns of different tissues, the mitochondrial enzyme always existed as a single component which had the same electrophoretic mobility irrespective of the tissue of origin. Wood (1963) has purified this enzyme from ox brain and has shown that it gave three peaks after chromatography on DEAE-Sephadex. One appeared to be due to association of the enzyme with other enzymes or proteins while each of the other two gave two peaks upon rechromatography, indicating the possibility that they represented interconvertible forms of the enzyme.

A. SUB-UNIT STRUCTURE

A sub-unit hypothesis has been proposed for the molecular structure of this enzyme. Of the three isoenzymes detected after agar-gel electrophoresis, the one with intermediate mobility was possibly a combination of sub-units from the two extreme isoenzymes (Burger *et al.*, 1964).

Experimental evidence for a dimeric structure for creatine kinase has recently been presented (Dawson *et al.*, 1965). By dissociating isoenzymes from chick brain, heart and muscle with 6·5 M guanidine or by freezing and thawing in the presence of salt and sodium phosphate, hybrid isoenzymes have been produced. The entity formed by dissociation and recombination of that from muscle with that from heart or brain had an electrophoretic mobility intermediate between those of the two parental types (Fig. 20). A naturally occurring enzyme with this mobility could also be detected after electrophoresis of chick heart extracts. The kinetic characteristics of the hybrid enzyme produced *in vitro* were also intermediate between those of the parental types. They were identical with the naturally occurring isoenzyme. It would appear that the

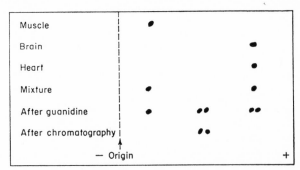

FIG. 20.   Dissociation and recombination of creatine kinase isoenzymes from chick tissues (reproduced with permission from Dawson *et al.*, 1965). After guanidine treatment, the brain and hybrid enzymes sometimes appear as doublets.

muscle isoenzyme is probably made up of two identical sub-units (M—M) and that from brain of two identical sub-units (B—B). Both the naturally occurring hybrid enzyme and that produced *in vitro* would then have the structure M—B (Dawson *et al.*, 1965).

4. PYRUVATE KINASE

Using starch block electrophoresis and subsequent elution of serial segments, differences in the electrophoretic mobility of the enzymes from human leucocytes and erythrocytes have been established (Koler *et al.*, 1964). The enzymes could not be separated by gel-filtration but were precipitated by different concentrations of ammonium sulphate.

After agar gel electrophoresis of extracts of rat organs, Fellenberg *et al.* (1963) have shown differences in the electrophoretic mobility of the enzymes from liver, kidney, muscle, heart and brain. Only one zone of activity could be detected in each tissue.

5. CARBONIC ANHYDRASE

Using chromatography on columns of DEAE cellulose and cellulose column electrophoresis, Lindskog (1960) has separated purified carbonic anhydrase from bovine erythrocytes into two components. Nyman (1961) has separated the human erythrocyte enzyme into two major components and could identify a further two minor components in some preparations. Chromatography on columns of hydroxyapatite has also resulted in the separation of the human erythrocyte enzyme into two peaks (Rieder and Weatherall, 1964). Each of these could be separated into a further two components by starch gel electrophoresis (Rickli *et al.*, 1964). Laurent *et al.* (1962) have described three zones of enzyme activity after electrophoresis and termed them X_1, X_2 and Y.

Using cellulose acetate electrophoresis Korhonen and Korhonen (1965) have shown that the rat enzyme existed in multiple forms in erythrocytes, brain, kidney and lens homogenates. Bovine lens carbonic anhydrase had previously been separated into two components (Sen *et al.*, 1963) and the rat enzyme apparently showed a similar pattern. The mobilities of the two faster moving zones of the rat erythrocyte enzyme were identical with those of the lens, brain and kidney.

Tappan *et al.* (1964) have compared the enzymes from bovine, guinea-pig and human erythrocytes by acrylamide disc electrophoresis and DEAE-cellulose column chromatography. They found that the isoenzyme patterns differed from species to species.

D

CHAPTER IV

Isoenzymes of the Hydrolases

1. Non-specific alkaline phosphatase

The existence of serum alkaline phosphatase in more than one form was first indicated by paper electrophoresis (Baker and Pellegrino, 1954; Keiding, 1959; Rosenberg, 1959). After starch-gel electrophoresis, activity was extracted from two zones, one of which moved more slowly than the slow α_2-globulin and the other slightly more slowly than the β-globulin (Kowlessar *et al.*, 1958). Both zones have shown increased activity in liver disorders but there has been an increase in the β-globulin region only in bone disease. The test paper method for visualizing the zones of activity was soon introduced (Estborn, 1959). This has demonstrated a major band which travelled slightly more slowly than β-globulin and a second faint band corresponding to what has been described as the prealbumin (acid α_1-glycoprotein) zone. A third zone of activity near the origin has also been demonstrated in bile.

Attention was now concentrated on the actual staining of regions of activity within the gel. With a discontinuous buffer system and vertical gel electrophoresis, bands of activity have been visualized by means of a substrate containing α-naphthyl phosphate and by staining the liberated naphthol with Fast Blue RR (Boyer, 1961). Some sixteen bands of alkaline phosphatase have been detected in all the different human sera examined. All the bands have never been demonstrated in a single individual. It has been claimed that they occur in groups and that among over 700 sera investigated none have shown more than four groups or have had more than eight distinct bands. The groups were labelled A, B, C, D, E and F; normal adults have been found to have one or two C components and occasionally a faint band in the F zone. Certain of the groups, viz. A, B and D, have been observed only in pregnancy; the last named has been limited to specimens obtained from negresses.

Following starch-gel electrophoresis in 0·05 M Tris-HCl buffer at pH 8·6, a staining technique has been elaborated which makes use of Ca-α-naphthyl phosphate and Brentamine Fast Red TR (Hodson *et al.*, 1961; 1962). The alkaline phosphatase activity demonstrable in human serum in liver disease moved with a different mobility from that present in bone disease and more than one band has been shown to occur both in tissue extracts and in pathological sera. The patterns obtained with extracts of bone (costochondral junction), liver, intestine, kidney and placenta are shown in diagrammatic form in Fig. 21 alongside a slice of the gel stained for protein bands. In the β-lipoprotein region, minor bands are often found in fresh extracts of tissues; these tend to vanish on storage.

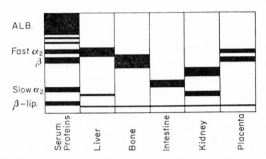

Fig. 21.   Alkaline phosphatase isoenzyme patterns visualized after starch gel electrophoresis of saline extracts of human tissues.

According to one group of investigators (Moss et al., 1961a) tissue alkaline phosphatases partially purified by a process of discontinuous starch-gel electrophoresis have so-called "K_m" values characteristic of the tissue of origin. These values have not actually been true Michaelis constants but have been determined by the assessment of pH optima at varying substrate concentrations. It has been shown (Moss and King, 1962) that a number of active alkaline phosphatase fractions could be extracted from segments of the gel after electrophoresis of concentrated butan-1-ol extracts of human bone, liver, kidney and small intestine. Each of the fractions from a given organ was said to have the same "K_m" value but there have been differences between fractions from different tissues. It was suggested that the different bands were possibly complexes with different proteins but actually fractions of the same enzyme. Recovery of activity from the β-lipoprotein region, by freezing and thawing followed by a second electrophoresis, fractionated the band into a portion moving with the original mobility together with a faster moving component. It was suggested that the latter could have resulted from the dissociation or degradation of a complex between the enzyme and the lipoprotein.

In the hands of many workers (Boyer, 1961; Chiandussi et al., 1962; Hodson et al., 1962; Moss and King, 1962) the main bands of alkaline phosphatase activity have appeared somewhat diffuse and it has long been felt that they were in fact heterogeneous. Using a butan-1-ol extract of human small intestinal mucosa, a single broad band of alkaline phosphatase has now been resolved into two components after an extended electrophoresis run in a discontinuous buffer system (Moss, 1963). During storage at $-20°C$ for some four months, the phosphatase pattern in intestinal mucosal extracts changed in so far as the fastest zone moved even faster and a third zone became apparent. At the same time the pattern obtained by chromatography on substituted cellulose columns became more complex, whereas with Sephadex G200 the tissue extract appeared as a single smooth peak (Moss, 1963). It has, therefore, been suggested that the different mobilities on starch gel are due to differences of charge rather than of molecular size. This does not seem to fit in with the concept that the different fractions of a tissue are complexes with different proteins, since it would be somewhat surprising if these were all of the same molecular weight.

Starch gel electrophoresis of butanol extracts of small intestine mucosa of young mice has shown four zones of phosphatase activity (Moog et al., 1966). One of these is prominent in extracts of the proximal duodenum and is present only in small amounts in extracts of distal duodenum and jejunum. Phosphatases of identical electrophoretic mobility from jejunal and proximal duodenum have, however, markedly different kinetic characteristics, e.g. relative reaction rates with β-glycerophosphate and phenyl phosphate. Human kidney alkaline phosphatase has been separated into three distinct zones by starch gel electrophoresis and into four fractions by chromatography on DEAE-Sephadex (Butterworth and Moss, 1966). Each of the chromatographic peaks had a different mobility during starch gel electrophoresis but after treatment with neuraminidase they all had identical mobility. It would appear, therefore, that the multiple components of kidney alkaline phosphatase may differ only in their content of varying amounts of bound sialic acid.

Using chromatography on DEAE and CM-celluloses, Grossberg et al. (1961) have isolated highly active phosphatase preparations from human liver, bone, kidney, spleen and intestine. Although three or four components could be seen in kidney and intestinal extracts, the phosphatases from other tissues could not be resolved into more than one fraction.

Two fractions of alkaline phosphatase have been separated from sheep brain, using chromatography on DEAE cellulose (Saraswathi and Bachhawat, 1966). These two isoenzymes had similar pH optima and substrate specificities but one of them had a much higher affinity towards pyridoxal phosphate and adenosine diphosphate. This isoenzyme predominated in the white matter of brain.

Alkaline phosphatase obtained from E. coli (see Chapter VIII) gave rise to several enzymatically active bands after zone electrophoresis

(Bach *et al.*, 1961). Mutant forms also showed similar isoenzyme constituents with somewhat different chemical structure.

The alkaline phosphatase activity of a number of human tissue extracts such as liver, bone, kidney, uterus, ovary, breast, pancreas, muscle, brain, heart, lung, prostate and testis has been reduced by half after heating at 56°C for 30 minutes (Moss and King, 1962; McMaster *et al.*, 1964; Neale *et al.*, 1965). Placental alkaline phosphatase, however, was not inactivated to any significant extent by heating below 75°C in the presence of 10 mM magnesium sulphate (Neale *et al.*, 1965).

A. IMMUNOLOGICAL STUDIES

Schlamowitz prepared a rabbit antiserum to a preparation of dog intestinal phosphatase which precipitated the intestinal phosphatase from solution (Schlamowitz, 1954a) but not the phosphatases from dog liver or dog kidney, nor those from rabbit, rat or bovine intestine (Schlamowitz, 1954b). Antisera have also been prepared in rabbits to the alkaline phosphatases from human intestine and from human osteogenic sarcomatous bone (Schlamowitz and Bodansky, 1959). The enzymes from human intestine and bone could be differentiated using these antisera. In the presence of horse anti-rabbit γ-globulin, the anti-bone phosphatase serum gave cross reactions between the enzymes from intestine, kidney and liver (Schlamowitz and Bodansky, 1959).

Boyer (1963) has prepared rabbit antisera to the alkaline phosphatases from human liver, kidney, bone, intestine and placenta. Using starch gel electrophoresis of supernatant fluids after precipitation by anti-enzyme sera, he has shown the presence of three antigenic classes of alkaline phosphatase. The enzymes from bone, liver, spleen and the major kidney phosphatase component constituted the first class and those from intestine and placenta formed the second and third classes respectively. It has been found that the second and third class enzymes showed cross reaction amongst themselves and with the minor kidney phosphatase. This latter would indicate that there are antigenic differences in the isoenzyme fractions of human kidney extract.

B. PHYSIOLOGICAL CONSIDERATIONS

For a considerable time it has been assumed that human serum alkaline phosphatase is of skeletal origin (Armstrong and Banting, 1935; Gutman and Jones, 1949). This concept has apparently been supported by immunological evidence (Schlamowitz, 1958; Schlamowitz and Bodansky, 1959). Isoenzyme studies using starch-gel electrophoresis have, however, demonstrated that in normal human adult serum the major alkaline phosphatase constituent is derived from the liver (Hodson *et al.*, 1962; Cunningham and Rimer, 1963). This component is constantly present in serum but other components may also be detected, although not constantly, which correspond to intestinal and bone alkaline phosphatase. It is interesting to note that the latter is apparently the least common constituent. Support for the derivation of the former band from intestine was apparent when Fishman and Kreischer (1963) showed that the slower moving normal serum alkaline phosphatase was inhibited by L-phenylalanine which is known to inhibit human intestinal alkaline phosphatase. This finding was confirmed by Robinson and Pierce (1964) who also showed that this second alkaline phosphatase was resistant to neuraminidase, whereas the removal of neuraminic acid groups from the faster moving alkaline phosphatase produced a decreased mobility of the enzyme (Fig. 22).

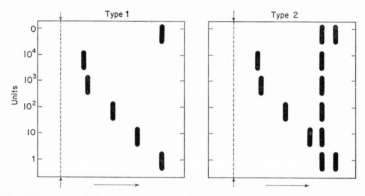

FIG. 22.   Effect of varied amounts (arbitrary units) of neuraminidase on human serum alkaline phosphatases. The mobility of the additional component of the Type 2 phosphatase is not affected by neuraminidase (reproduced with permission from Robinson and Pierce, 1964).

The presence of the second zone has been confirmed by other workers (Arfors *et al.*, 1963a; 1963b; Robinson and Pierce, 1964; Bamford *et al.*, 1965; Schreffler, 1965). Population studies have indicated that this slow-moving alkaline phosphatase is under genetic control and a relationship with the ABO blood group system has been established (Arfors *et al.*, 1963a; 1963b; Bamford *et al.*, 1965; Schreffler, 1965).

Fishman and co-workers (1963) have been able to show that L-phenylalanine is a stereo-specific tissue specific inhibitor for the alkaline phosphatase of human intestine. Estimation of the alkaline phosphatase activity of extracts of human liver and intestine at pH 9·8 in veronal buffer has shown that the enzyme from human liver was inhibited by about 10% in 0·5 mM L-phenylalanine, whereas the enzyme from human intestine was inhibited by about 80%. Using normal human serum, the alkaline phosphatase activity at pH 9·8 has been reported as being reduced by some 20–60% in 5 mM L-phenylalanine with a mean value of 40% (Fishman *et al.*, 1965). Using phenyl phosphate or *p*-nitrophenyl phosphate as substrate in bicarbonate buffer at pH 10·2, Keiding (1966) has shown that the normal serum alkaline phosphatase activity was reduced by 10% in 5 mM L-phenylalanine, whereas the enzyme from intestinal fluid was reduced by some 40% under the same conditions. The presence of 3·4 mM magnesium chloride in the buffered substrate has been shown to reduce the extent of the L-phenylalanine inhibition of normal serum (Keiding, 1966).

It has been reported that those individuals with the two zones of serum alkaline phosphatase have significantly higher total enzyme activities than those with a single zone (Bamford *et al.*, 1965). Fishman and co-workers (1965) have also found significantly higher levels of serum phenylalanine sensitive phosphatase in ambulatory patients than in hospitalized patients and have shown that the higher activity is of intestinal origin.

In the serum of children, as one would expect, the major alkaline phosphatase activity obtained after starch-gel electrophoresis corresponded to that of bone (Taswell and Jeffers, 1963). This is, of course, related to the increase in osteoblastic activity during the growth period. Beckman and Grivea (1965) have also shown that the serum alkaline phosphatase of the newborn had a slower mobility than the enzyme from normal adults.

Using electrophoresis on Pevikon C870, a copolymer of polyvinyl chloride and polyvinyl acetate, Nordentoft-Jensen (1964) has shown that the serum alkaline phosphatase from normal children and from adult patients with bone disease has a slower mobility than the serum enzyme from patients with liver disease.

Characterization of human tissue and serum alkaline phosphatases using electrophoretic mobility, heat stability and phenylalanine inhibition as parameters has further confirmed that the adult serum phosphatase is of hepatic origin and that the serum phosphatase of children is a mixture of the bone and liver phosphatases (Warnock, 1966). The serum alkaline phosphatase which is considered to be of intestinal origin has been detected in some children and adults.

Fractionation of adult human serum alkaline phosphatase by separation on Sephadex G200 has also indicated that the main phosphatase activity was derived from the liver and that the alkaline phosphatase of bile behaved similarly during gel filtration to the alkaline phosphatase from bone (Estborn, 1964). This has been taken to support the concept that bone phosphatase is usually excreted through the bile.

The protein characteristics of serum and bile alkaline phosphatase have been examined using starch gel electrophoresis and preparative ultracentrifugation (Pope and Cooperband, 1966). The results indicated that the enzymes from bile and serum were similar in molecular size but had different isoenzyme patterns after starch-gel electrophoresis.

Using starch block electrophoresis, Keiding (1964) has reported on the alkaline phosphatase present in human lymph. This could be separated into major and minor fractions. The former has been shown to have the same mobility as the intestinal enzyme. There has been an increase in this major component in lymph obtained by cannulation of the thoracic duct after fat ingestion. A serum phosphatase with the same mobility as the main lymph phosphatase has been detected in only four out of 250 sera examined (Keiding, 1964). This is difficult to understand in view of the postulated identity with intestinal alkaline phosphatase. It has recently been found that the lymphatic and intestinal phosphatases show similar inhibition by 5 mM phenylalanine as well as similar substrate affinities (Keiding, 1966).

With agar-gel electrophoresis only one band of activity has been demonstrated in adult serum but its position corresponded to one of the two bands which have been obtained from liver extracts (Haije and de Jong, 1963).

Using gel filtration on Sephadex G200 and chromatography on DEAE-Sephadex, three types of ovine serum alkaline phosphatase patterns have been isolated (Aalund et al., 1965). There was no evidence for a complex between any of the serum alkaline phosphatase isoenzymes and blood group O substance.

As has already been mentioned, certain specific changes in the serum alkaline phosphatase pattern obtained by starch-gel electrophoresis occur in pregnancy (Boyer, 1961). It has also been demonstrated that an alkaline phosphatase pattern identical with that obtained from placental extracts appears in the serum during the last six weeks of pregnancy (Latner, 1965). It is assumed that this pattern is due to the liberation of enzyme into the circulation as the result of placental degeneration. Other workers have suggested that serum alkaline phosphatase in pregnancy is obtained from bone (Meade and Rosalki, 1963).

Two types of serum alkaline phosphatase patterns have been described in pregnant Swedish women (Beckman and Grivea, 1965). The patterns would appear to be similar to those described by Boyer (1961).

Using a visual staining technique after agarose gel electrophoresis, Takahashi et al. (1963) have shown that the serum alkaline phosphatase from pregnant women had a mobility intermediate between the α_2- and β-globulins, whereas the enzymes from adult males and children migrated as $\alpha_1\alpha_2$- and α_2-globulins respectively. Also using agarose as an electrophoresis medium, Dymling (1966) has found two components of serum alkaline phosphatase in pregnant women. One of these phosphatases had the same mobility as the phosphatase in placental extracts.

Studies of the heat stability of serum alkaline phosphatase during the later stages of pregnancy have shown that approximately half the maternal serum phosphatase is probably of placental origin and that the foetal serum enzyme is not (Kitchener *et al.*, 1965). The electrophoretic mobility of the placental enzyme and the enzyme from maternal serum appeared to be slightly lower than that of normal males and non-pregnant women (Kitchener *et al.*, 1965).

Starch-gel electrophoresis of urinary alkaline phosphatase, using a discontinuous buffer system, has shown that the commonest pattern obtained was a single zone migrating further towards the anode than any of the phosphatases in tissue extracts (Butterworth *et al.*, 1965). Chromatography on Sephadex G200 has indicated that the urinary phosphatase was of smaller molecular size than kidney alkaline phosphatase. Evidence for the renal origin of this urinary isoenzyme has been the detection of an alkaline phosphatase with the same mobility as the urinary component in the cells shed from the kidney tubules after aspirin administration.

It has long been recognized that alkaline phosphatase occurs in dental pulp. Although it has been tempting to identify it with that of bone, no supporting evidence was obtained until the demonstration that the major activity of extracts of dental pulps appeared in the same position as that from costochondral junction (Hodson *et al.*, 1965). Whilst this is by no means certain evidence, it is at least highly suggestive that odontoblasts produce an alkaline phosphatase which is the same as that produced by osteoblasts.

2. Specific alkaline phosphatases

Specific alkaline phosphatases have been demonstrated in a study of Golgi-body associated phosphatases (Allen, J. M., 1963; Allen and Hynick, 1963). Nucleoside diphosphatase and thiamine pyrophosphatase have been separated in acrylamide gels. Thiamine pyrophosphatase reacts most strongly with thiamine phosphate but also gives weak reactions with cytidine, guanosine and inosine diphosphates. Nucleoside diphosphatase reacts strongly with uridine and inosine diphosphate and gives weak reactions with guanosine, thiamine and cytidine diphosphates. Little or no reaction has been obtained with either enzyme on

D*

adenosine diphosphate (Allen, J. M., 1963). The specific substrates indicated that each band represented a different enzyme. With the other substrates, the bands would appear to have been isoenzymes. This illustrates a real difficulty in terminology.

Starch-gel electrophoresis of rat liver extract has shown six separate glycerophosphatases (Sandler and Bourne, 1961; 1962). The alizarin method for specific alkaline phosphatases has demonstrated bands which hydrolysed AMP and ATP, as well as glycerophosphate. Other bands were specific for NAD, NADP, fructose-6-phosphate, glucose-6-phosphate and hexose diphosphate. The same workers have shown that a commercial preparation of calf intestinal alkaline phosphatase gave only one band after starch-gel electrophoresis. This was capable of hydrolysing AMP, ADP, ATP, NAD, NADP, creatine phosphate, α-glycerophosphate, glucose-6-phosphate, fructose-6-phosphate, glucose-1-phosphate and acetyl phosphate. Ulrich (1964) has presented evidence for multiple mitochondrial adenosine triphosphatases with different pH optima and different nucleotide specificities.

3. ACID PHOSPHATASE

Acid phosphatases are present in most tissues and body fluids and they can be differentiated by a number of physicochemical techniques. In 1934, Davies was able to show that the enzyme from erythrocytes hydrolysed α-glycerophosphate more readily than β-glycerophosphate, whereas that from spleen was more effective with β-glycerophosphate. Herbert (1944) demonstrated that prostatic acid phosphatase was irreversibly inactivated by incubation with 40% ethanol for 30 minutes at room temperature. Alcohol inhibition differentiated between prostatic and other acid phosphatases in pathological human sera (Herbert, 1945; 1946). Some doubts as to the specificity of this method of inhibition have been raised by King and co-workers (1945), who have shown that the erythrocyte acid phosphatase is also inhibited by prolonged incubation with ethanol. Further inactivation studies have indicated that the enzymes from human adrenal, intestine, liver, pancreas, spleen and thyroid were not affected by ethanol, whereas the enzymes from bile, kidney, erythrocytes and prostate were markedly inhibited (Abdul-Fadl and King, 1947).

Differentiation between erythrocyte and prostatic acid phosphatases has been obtained by using formaldehyde or tartrate. Formaldehyde was a potent inhibitor of the acid phosphatase from erythrocytes but did not affect that from prostate (Abdul-Fadl and King, 1947) while L-tartrate was found to be a potent inhibitor of prostatic acid phosphatase but did not affect the enzyme from erythrocytes (Abdul-Fadl and King, 1948). D-Tartrate inhibition of tissue acid phosphatases appeared to be pH dependent; the enzymes from liver, kidney and spleen were inhibited in the acid range pH 3·8–4·6 but not in an alkaline medium. The enzyme from erythrocytes appeared to be inhibited below pH 4·8 but was slightly activated above this pH (Abdul-Fadl and King, 1948). This suggested the presence of two erythrocyte acid phosphatases, the existence of which had previously been reported by Roche et al. (1942). The effect of various inorganic ions on different acid phosphatases has been examined (Abdul-Fadl and King, 1949); copper ions inhibited erythrocyte acid phosphatase while ferric ions in the presence of acetate buffer inactivated the prostate enzyme.

Early investigations with starch-gel electrophoresis (Estborn, 1959; 1961; Estborn and Swedin, 1959) did not result in the resolution into more than one band of acid phosphatase activity in either human seminal plasma or serum. The procedure was, however, carried out at pH 8·9 and it is well known that prostatic acid phosphatase is very unstable at such high pH values. This would mean that even if resolution had occurred into two or more fractions, the minor bands might have been totally destroyed. Nevertheless, three fractions have been demonstrated after starch-gel electrophoresis of human serum at pH 8·6 (Dubbs et al., 1960). Using buffers of relatively low pH values, it has been possible by discontinuous electrophoresis in starch gel to demonstrate three bands of activity of human prostatic acid phosphatase (Sur et al., 1962). All three bands have not been inhibited by formaldehyde and so none could have been obtained from red blood cells. The Michaelis constants of the fractions differed from each other and the slowest moving component possessed the highest value. The fastest moving band appeared very broad and it has been possible to sub-divide this further. It was finally concluded that prostatic acid phosphatase was separable into at least thirteen and probably more bands by starch-gel electrophoresis in citrate buffer at pH 6·2. The multiple bands were apparently not artefacts. Up to seventeen separate acid phosphatase components could be identified in extracts of human liver, kidney and prostate (Lundin and Allison, 1966). Similar patterns could be identified in extracts of liver from mice, rats, rabbits and hamsters but extracts of mouse and rat kidney showed only three components. Acid phosphatase from human red cell haemolysates also showed more than one zone of activity after starch-gel electrophoresis and their genetic variants have been demonstrated (Hopkinson et al., 1963). Gel filtration on Sephadex G200 has shown that erythrocytes contain at least two acid phosphatases both of which are distinct from prostatic acid phosphatase (Estborn, 1964).

A number of acid phosphatases of rat liver have been separated on acrylamide gel. It is suggested that these represented a family of different enzymes, although some could be more closely related than others (Barka, 1961).

The intracellular distribution of acid phosphatase isoenzymes in rat and human liver has been investigated (Reith and Schmidt, 1964). Four isoenzymes could be detected in isolated parenchymal cells and different isoenzymes were associated with the particulate and supernatant fractions. Studies of the heat stabilities, pH optima and sensitivity to tartrate and fluoride of the cytoplasmic and lysosomal acid phosphatases of rat liver has shown the existence of two different forms of the enzyme (Nelson, 1966).

Four zones of acid phosphatase have been detected after gel electrophoresis of extracts of human brain; three zones were present in extracts of feline spinal cord (Barron *et al.*, 1964).

4. ESTERASES

In general, three groups of esterases have been shown to be present in tissues and body fluids. These are the arylesterases or A (aromatic) esterases, the aliesterases or B esterases and the cholinesterases (Aldridge, 1953a; 1953b; 1954; Augustinsson, 1958; 1959; 1961). The A, or arylesterases, hydrolyse aromatic esters more readily than aliphatic esters, while the B, or aliesterases, hydrolyse the latter esters more readily than do the A group. Cholinesterases, which are the third group, are most effective with choline esters and are also capable of hydrolysing aliphatic esters and aromatic esters but less easily (Augustinsson, 1961). The A-esterases of many sera have been shown to hydrolyse p-nitrophenyl acetate at a higher rate than the corresponding butyrate, while the B-esterases hydrolyse both esters at about the same rate (Aldridge, 1953a). The A and B esterases can also be differentiated by their sensitivity to organo-phosphorus inhibitors, the B esterases are very sensitive and the A esterases relatively unaffected (Aldridge, 1953a; 1954). Cholinesterases can be distinguished as they are inhibited completely by 10^{-5} M eserine (Richter and Croft, 1942; Augustinsson, 1959). Certain other esterases could be inhibited by higher concentrations of eserine (Augustinsson, 1959).

The use of starch gel electrophoresis has enabled multiple forms from animal tissues and sera to be demonstrated and characterized (Hunter and Markert, 1957; Markert and Hunter, 1959). Esterases of thirty-two mouse organs have been defined (Markert and Hunter, 1959). Using mouse liver as an enzyme source, the substrate specificities of some thirty esterase components have been studied. The esterases of mouse serum have been separated into nine components and the effect of nine substrates and five inhibitors investigated (Hunter and Strachan, 1961). The results indicated that the esterases present in mouse blood could not be readily classified into sub-groups warranting their classification into isoenzymes. The serum esterases of other species, including the monkey, cat, rabbit, rat and human being, have been investigated by similar methods (Lawrence *et al.*, 1960; Hess *et al.*, 1963). The substrates employed have been α- and β-naphthyl acetate, α-naphthyl propionate, α-naphthyl butyrate, β-naphthyl laurate, β-naphthyl myristate and naphthyl-AS-acetate. The inhibitors included 10^{-4} M and 10^{-5} M eserine, 10^{-5} M diisopropylfluorophosphate (DFP), 10^{-6} M myletase and 10^{-6} M diethyl-p-nitrophenylphosphate. The esterase patterns differed markedly between species and this suggested that the differences could be used to identify species and study genetic variations within a species (Hess *et al.*, 1963).

Multiple forms of non-specific esterase have been described after starch-gel electrophoresis of aqueous extracts of human liver (Ecobichan and Kalow, 1961; 1962; Ecobichan, 1965), skeletal muscle (Ecobichan and Kalow, 1965), kidney (Ecobichan and Kalow, 1964) and brain (Ecobichan, 1966b). The preliminary results on human liver have indicated that in regard to electrophoretic migration and enzymatic properties none of the liver esterases were identical with any of the serum esterases (Ecobichan and Kalow, 1961). In extracts of human kidney five zones of esterase activity were characteristic of serum cholinesterase and were believed to be due to serum contained in the renal tissue (Ecobichan and Kalow, 1964). A zone of isoenzymic esterases in renal and hepatic tissue had the properties of an acetyl esterase and an esterase component with the mobility of albumin was shown to be an aliesterase. Three discrete zones of esterase activity with affinities for α-naphthyl butyrate appeared to be characteristic of renal tissue (Ecobichan and Kalow, 1964). Fifteen zones of esterase have been demonstrated in extracts of human brain. Four of these bands were common to liver, muscle and kidney but have not been detected in serum (Ecobichan, 1966b). The majority of the brain esterases have shown non-specific behaviour towards various esters and inhibitors and it has been concluded that human tissues possess a heterogeneous mixture of carboxylesterases with a wide range of properties. Using starch-gel electrophoresis, other studies have been made of the separation and properties of human brain esterases (Barron *et al.*, 1961; Barron *et al.*, 1963), and rat brain esterases (Eranko *et al.*, 1962). At least eighteen bands with activity against naphthol esters have been found and, with the exception of the esterases hydrolysing α-naphthyl propionate, α-naphthyl butyrate or thiocholine esters, no differences have been obtained in different regions of the human brain. The number of human esterases has been found to be considerably greater than in the rat, rabbit, cat or guinea pig, but the authors have not indicated a definite relationship between esterase multiplicity and the higher organization of the human central nervous system. The effect of perfusion of organophosphorus compounds which inhibited serum esterases has been used to show that contained blood did not contribute towards the esterase patterns observed. Starch-gel electrophoresis has been used for characterization of desmo- and lyo-esterases in the sympathetic and spinal ganglia of the rat (Eranko *et al.*, 1964).

Further studies with human liver esterases have shown three areas of enzyme activity after starch-gel electrophoresis, with each made up of several zones of enzyme activity (Ecobichan, 1965). The relative molecular sizes of some of the esterase components have been compared by studying the retardation when the starch concentration was increased and in general it has been found that many of the multiple forms of esterase within a particular group had similar molecular weights and thus differed primarily in net charge (Ecobichan, 1965). Variation in the mobilities of some of the esterase components of adult human liver have been detected (Ecobichan and Kalow, 1961; Blanco and Zinkham, 1966) but genetic control of this variation has yet to be established.

In a study of the esterase and phosphatase isoenzyme patterns of the human gastro-intestinal tract in the normal state and in non-tropical sprue, Weiser and co-workers (1964) have been able to confirm the finding of Markert and Hunter (1959) that there were specific esterase patterns for different parts of the gastro-intestinal tract.

Six components of the enzyme which hydrolyses α-naphthyl acetate have usually been demonstrated after agar gel electrophoresis of human tissue extracts (Oort and Willighagen, 1961). Although different tissues normally gave different esterase patterns, a variety of types of carcinoma of the lung all gave identical patterns.

Micheli and Grabar (1961) have examined the esterase patterns after electrophoresis of human haemolysates. Of the four zones of esterase activity detected, none could be identified as cholinesterase which appeared to be associated with the cell stroma.

A comparative study of the liver, lung and kidney esterases of some twenty species has been made with starch gel electrophoresis (Coutinho *et al.*, 1965). The animals investigated included such species as the oppossum, armadillo, whale, goat and squirrel. Both starch and acrylamide gel electrophoresis have been employed in an investigation of esterases in the Crustacean nervous system (Maynard, 1964). The non-specific esterases exhibited zymogram patterns which were relatively tissue and species specific for the lobsters, *Panularis argus, Panularis guttatus* and *Homarus americanus*. Two acetylcholinesterases moved anodically at pH 8·9, the more rapidly migrating enzyme was found to be more predominant in the central ganglia and interganglionic connectors whereas the slower was more predominant in peripheral nerves. Meier, Jordan and Hoag (1962) have used agar gel electrophoresis to study esterases in relation to genotypic differences. Aberrant patterns of esterases and cholinesterases were detected in tissues of some mice with neuromuscular mutations.

Blanco and Zinkham (1966) have investigated the changes in soluble esterases during development of human tissues. These workers have described a fast-migrating group of three esterases after starch gel electrophoresis and have shown that their intensity varied from tissue to tissue, being most marked in the kidney and adrenal gland. All tissues possessed a slow-moving group; up to six zones of varying intensity could be detected in extracts of human liver, stomach, diaphragm, muscle, heart, testis, adrenal and kidney. A third group of up to five esterases of intermediate mobility has been described in some tissue extracts (Blanco and Zinkham, 1966). The fastest moving groups have been shown to be sensitive to 10^{-4} M eserine. Maturation from the foetal to neonatal and adult stages was shown to be accompanied by changes in the patterns; in general the number and intensity of the esterases increased during development. These results were contrary to those reported by Paul and Fottrell (1961), who found no differences in the esterase zymograms from foetal and adult human tissues, although ontogenic changes in the esterase patterns of mouse tissues have been described (Markert and Hunter, 1959).

Zymograms of denervated feline muscle have shown a redistribution of activity of some of the aliesterase components and it has been postulated that these changes may have some specificity for neural atrophy in so far as the muscle-wasting following tenotomy did not appear to be accompanied by increases in any of these isoenzyme fractions (Barron *et al.*, 1966). These workers have also described an increase in aliesterase activity following denervation but have found no consistent abnormality in the cholinesterase isoenzyme patterns of denervated tissue. Although changes in the non-specific esterase isoenzyme patterns similar to those in denervated tissue have not been observed, a striking activation of the cholinesterase isoenzymes could be detected following tenotomy (Barron *et al.*, 1966).

Changes in the esterases of rat brain and blood serum following acute cranial exposure to X-rays have been investigated (Masurovsky and

Noback, 1963). Acrylamide gel electrophoresis has indicated marked alterations. The most profound have been in the blood, in which several esterase components have been decidedly inactivated by irradiation. The brain tissue patterns have shown much less marked changes in the activity of certain components.

Investigations have been carried out using cultured cell strains from different species grown in the same environment for some years to see whether the differences in esterases among organs and species might be the result of environmental rather than genetic factors (Paul and Fottrell, 1961). All human cell lines have exhibited a typical human esterase pattern distinct from mouse cells. No alteration in pattern has been obtained by growing the cells in high concentrations of acetylsalicylic acid. HeLa, WISH and RA amnion cell lines all had similar esterase patterns (Beckman and Regan, 1964). There was little variation in the esterase pattern of fresh tissues and cells cultured for many years. Further studies of normal human cells and tumour cells grown *in vitro* have shown the presence of up to seventeen non-specific carboxylic esterases with eight or more sub-groups characterized on the basis of their interaction with activators, temperature stability and substrate affinities (Komma, 1963). The characteristic enzyme patterns found in the cultured cells might provide information as to their cellular function and differentiation, although there are at present no reports on changes in esterase patterns during culture.

Using electrophoresis on cellulose columns, extensive studies of vertebrate plasma esterases have also revealed three types of groups, viz. aryl esterases designated (ArE), aliesterases (AliE) and cholinesterases (ChE). Some plasmas contained all three esterase types and others only one or two types (Augustinsson, 1959; 1961). Most mammalian plasmas contained multiple aryl esterases. Human plasma was shown to have two forms which differed in their heat stability and sensitivity to lanthanon ions. Aliesterase has been found to be absent in the human but was the main esterase in the plasmas of lower vertebrates.

The enzymes from different species have been shown to possess varying electrophoretic mobilities, although substrate specificities were similar (Augustinsson, 1961).

The existence of these groups of esterases does not in itself mean that multiple forms of esterases or esterase isoenzymes exist, as esterases have a relatively wide range of substrate specificities. Species specificity of the esterases has been described in so far as most mammals possess a plasma acetyl-arylesterase and a butyrylcholinesterase while fish and bird plasmas possess a propionyl aliesterase and low concentration of acetylcholinesterase.

Cholinesterases have been considered as a separate group because of the relatively high hydrolysis rate obtained with choline esters and their high sensitivity to eserine. Tissue- and species-specific forms of cholinesterases have been detected (Augustinsson, 1958; 1959; 1961), although only a single fraction could be found in human plasma with mobility between α_2- and β-globulins.

Electrophoresis of human serum arylesterases on cellulose columns at pH 8·0 has shown the presence of three peaks of activity (Wilde and Kekwick, 1964). The first form moved ahead of the albumin and showed a preference for acetylesters, the second form was associated with the albumin and hydrolysed butyrylesters most readily, while the third peak was cholinesterase.

Two different types of cholinesterase in man and other vertebrates were first described by Alles and Hawes (1940), who showed that true cholinesterase hydrolysed acetylcholine most rapidly, whereas pseudocholinesterase was more effective with longer chain choline esters such as butyrylcholine. True cholinesterase is present in high concentrations in nerve and muscle tissue and erythrocytes 'pseudocholinesterase' in plasma, liver and brain white matter. It has been suggested that, as high concentrations of acetylcholine and other esters inhibited true cholinesterase, a possible function of 'pseudocholinesterase' is to protect the true cholinesterase by hydrolysing the substrate inhibitors (Lehmann and Silk, 1953).

Evidence for the electrophoretic heterogeneity of human serum cholinesterase was first obtained by Pinter (1957), de Grouchy (1958) and Dubbs and co-workers (1960). The latter group of workers have been able to demonstrate that after starch-gel electrophoresis of serum at pH 8·6 two zones of cholinesterase could be detected in the globulin region; both zones were inhibited by eserine. Single zones of cholinesterase have been reported (Ecobichan and Kalow, 1961; Hunter et al., 1961; Thompson and Cook, 1961) but the majority of observations have indicated the presence of three or four minor zones of this enzyme (de Grouchy, 1958; Lawrence et al., 1960; Bernsohn et al., 1961; Paul and Fottrell, 1961; Uriel, 1961; Harris et al., 1962; Latner, 1962; Stern and Lewis, 1962; Hess et al., 1963; Hunter et al., 1964; LaMotta et al., 1965; Dubbs, 1966). Other vertebrate sera also contain multiple forms of cholinesterase (Lawrence et al., 1960; Bernsohn et al., 1961; Hunter and Strachan, 1961; Hunter et al., 1961; Paul and Fottrell, 1961; Hess et al., 1963; Kaminski and Gajos, 1964; Oki et al., 1964; Reiner et al., 1965).

Using two-dimensional starch gel electrophoresis of human serum, Harris et al. (1962) have been able to demonstrate four areas of enzyme activity by the hydrolysis of α-naphthyl acetate. All four components were inhibited by physostigmine.

A concentrated preparation of normal human serum cholinesterase has been separated into ten components by starch-gel electrophoresis in a discontinuous buffer system (Lehmann and Liddell, 1964). Only five zones have, however, been obtained with a purified concentrate of serum cholinesterase (LaMotta et al., 1965). When these were eluted from the gel and concentrated by precipitation with 65% ammonium sulphate, each of the five fractions had the mobility of the originally slowest component; elution from the gel and re-electrophoresis without concentration gave an isoenzyme pattern in which a portion of each isoenzyme retained its original mobility and the remainder was converted into the fastest moving isoenzyme. This indicated that the five isoenzymes were interconvertible and were made up of polymers. Each of the four faster moving isoenzymes formed the major slowest moving component in the absence of the other components. This suggested a stepwise pathway for polymerization (LaMotta et al., 1965). Evidence for different molecular weights of the isoenzymes has also been derived from gel filtration studies on Sephadex G200 (Harris and Robson, 1963).

Alterations in the electrophoretic mobility of human serum cholinesterase after incubation with neuraminidase have been described (Svensmark, 1961a; 1961b) and some abnormal mobilities of the enzyme in biological fluids have been attributed to variations in the number of sialic acid residues attached to the enzyme; the number of such residues being dependent on the neuraminidase content of the fluid in question.

The number and intensity of the serum cholinesterase components could be increased by subjecting serum to treatment with ultrasonics (Dubbs, 1966). It is possible that the extra zone which was detectable after such treatment could represent enzyme previously bound to the β-lipoprotein moiety.

Uriel (1961) has described four zones of a human serum esterase which hydrolyses α-naphthyl acetate; the one with the least mobility being classed as a cholinesterase, due to its inhibition by 2×10^{-5} M eserine. In rat serum, three components have been detected; two in the pre-albumin region and one in the α_1-globulin region (Hermann et al., 1963). Mouse serum also showed four such esterases, one each in the pre-albumin, albumin, fast-α_2- and slow-α_2-globulin fractions (Talal et al., 1963). In mouse urine two esterase components have been detected, one of which had the same mobility as a kidney esterase (Talal et al., 1963). In human urine similar patterns were present and some of the urinary esterase appeared to be of renal origin (de Vaux St. Cyr *et al.*, 1963). Hirschfeld (1960) has described individual variations in the mobility of a human serum α_2-globulin which hydrolyses indoxyl acetate. Of the twelve sera examined, three types could be demonstrated according to whether the electrophoretic mobility of the enzyme was fast, slow or intermediate. Over half of those examined were of the intermediate type.

The effect of fat ingestion on the esterase isoenzyme patterns of rat intestine, intestinal lymph and serum has been studied (Lewis and Hunter, 1966). A large increase in two of the faster moving α-naphthyl butyrate hydrolases could be detected in the intestinal lymph within two hours of a meal of corn-oil. This increase in specific isoenzymes could also be detected in the serum but not as soon after the meal. The fat meal had no effect on the esterase activity of bile although bile was essential for the increased lymphatic esterase activity. It has been suggested that the source of the increased activity was most probably the small intestine and duodenum (Lewis and Hunter, 1966).

Immunoelectrophoretic studies of the serum esterases of the duck and the quail have been described (Kaminski, 1966). Sera from other *Gallinacea* were used for comparison in these studies, which indicated the presence of several esterase components, some of which were subject to inter- and intra-species variations. Changes in the serum esterases of adult and embryo chickens have also been examined using immunoelectrophoresis (Croisille, 1962).

Gel filtration of cow's milk on Sephadex G100 and G200 has been used to demonstrate four tributyrinases (Downey and Andrews, 1965). The same techniques have shown the presence of six such enzymes in extracts of rat adipose tissue.

5. ARYLAMIDASE

Using starch-gel electrophoresis and visual staining techniques, a single zone of activity of arylamidase has been detected in normal human serum (Dubbs *et al.*, 1960; 1961; Smith *et al.*, 1962; Smith and Rutenberg, 1963). Estimations of the enzyme activity in serial segments of a starch-gel following electrophoresis have yielded similar results (Kowlessar *et al.*, 1960; 1961). Normal serum arylamidase also moved as a single zone during electrophoresis on paper (Smith *et al.*, 1962) and cellulose acetate (Smith and Rutenberg, 1963; Meade and Rosalki, 1964). In maternal serum at term, a second zone could be detected which was apparently derived from the placenta (Kowlessar *et al.*, 1961; Smith and Rutenberg, 1963; Mead and Rosalki, 1964). A pattern similar to that present in maternal serum has also been observed in cord blood (Meade and Rosalki, 1964). Goebelsman and Beller (1965) have separated two arylamidases from the serum of pregnant women, using chromatography on Sephadex G200. One of the enzymes hydrolysed leucine-β-napthylamide and the other cystine-di-β-naphthylamide. The use of serum cystine aminopeptidase in the diagnosis of twins had previously been suggested by Miller et al. (1964), who found that there was a correlation between the incidence of twin pregnancy and the maternal serum enzyme activity during pregnancy.

Multiple zones of arylamidase have been visualized after starch-gel electrophoresis of extracts of human liver, pancreas and bile (Schobel and Wewalka, 1962) and the mobilities of the enzymes from human liver, kidney, placenta and pancreas varied during cellulose acetate electrophoresis (Smith and Rutenberg, 1963; Meade and Rosalki, 1964). An examination by starch-gel electrophoresis of human tissue arylamidases has provided evidence of tissue specific patterns (Smith and Rutenberg, 1966). Extracts of normal and cancer tissue from the liver, breast and lymph node gave similar patterns. Although the serum arylamidase was invariably inhibited by 10 mM L-methionine, the enzymes from tissue extracts were either inhibited, activated or unaffected (Smith and Rutenberg, 1966).

A similar study of rat tissue arylamidase has shown that extracts of liver, kidney, small intestine, prostate and skin all gave two isoenzyme bands, one of which remained at the origin (Monis, 1964; 1965). This study provided evidence to suggest that dermal fibroblasts were the source of the plasma enzyme and that renal proximal tubular cells were the source of the urinary enzyme.

Starch block and starch-gel electrophoresis have been used in a study of the arylamidases in rat serum and urine (Monis, 1964). A single zone of the enzyme has normally been detected in serum with two zones in urine, similar to kidney extracts. The serum band moved more rapidly than the faster zone of urine or kidney.

Using the zymogram technique and ion-exchange chromatography, characterization of the enzymes hydrolysing acyl naphthylamides has been attempted with mono- or dihalogen acyl naphthylamides (Hopsu and Glenner, 1964) or the trihalogen derivatives (Hopsu et al., 1965). Using starch-gel electrophoresis of a hog kidney acylase preparation, the enzymes hydrolysing chloroacetyl-L-leucine and chloroacetyl-β-naphthylamide have been separated. After DEAE cellulose chromatography, the same preparation yielded three fractions, one which hydrolysed leucyl-β-naphthylamide, chloroacetyl-L-leucine and chloroacetyl-β-naphthylamide, a second which hydrolysed Naphthol AS acetate and a third acetanilide (Hopsu and Glenner, 1964). Comparison of the patterns given by extracts of guinea pig and mouse kidney homogenates after starch-gel electrophoresis has shown the presence of multiple forms of the enzymes which hydrolysed Naphthol AS acetate and chloroacetyl-β-naphthylamide with different patterns for each enzyme for each animal. Similar observations with trihalogen acyl naphthylamides have shown similar phenomena (Hopsu et al., 1965).

Using DEAE cellulose chromatography, Patterson and co-workers (1963) have been able to separate three enzymes which hydrolysed leucyl-β-naphthylamide (LNA) and three enzymes which hydrolysed leucinamide (Leu NH$_2$) from extracts of mouse ascites carcinoma cells. These enzymes have actually been separated as four groups; the first group contained an enzyme which hydrolysed only leucinamide, the second group contained enzymes which hydrolysed both substrates, the third group contained enzymes which mainly hydrolysed LNA but did have a little activity towards Leu NH$_2$, while the fourth group hydrolysed only LNA.

Column electrophoresis was later applied to the problem of aminopeptidases and peptidases and it has been possible to resolve ten selected aminopeptidases, dipeptidases and arylamidases in extracts of rat liver (Patterson et al., 1965). Three arylamidases have been shown to be present in normal rat livers and a fourth faster moving enzyme has been detected in livers with early pre-neoplasia induced by the azo dye 3'-methyl-4-dimethyl azobenzene (Patterson et al., 1965). Following starch block electrophoresis of normal human serum, Nakagawa and Tsuji (1966) have found two fractions which hydrolysed leucinamide, one of which also hydrolysed leucyl-β-naphthylamide. A further two fractions which hydrolysed leucinamide could be found in sera from patients with hepatitis but no additional fractions of leucyl-β-naphthylamidase could be detected.

6. AMYLASE

Paper electrophoresis of human serum followed by amyloclastic detection of amylase activity in the various electrophoretic fractions has shown that the major portion of the normal serum enzyme is associated with the albumin fraction (McGeachin and Lewis, 1959; Dreiling et al., 1963). In patients with pancreatitis there appeared to be a second amylase fraction in the γ-globulin region. Baker and Pellegrino (1954) had previously found human serum amylase activity only in the γ-globulin fraction.

Electrophoresis of serum from the mouse (Delcourt and Delcourt, 1953) and the rat (McGeachin and Potter, 1961) has shown amylase activity in the β-globulin fraction. Three zones of amylolytic activity have been observed after paper electrophoresis of rabbit serum, one with the albumin, one with the γ-globulin and the third with a mobility between the β- and α_2-globulins (Berk et al., 1963). Elution of the protein fractions after electrophoresis, followed by estimation of saccharogenic activity, confirmed the presence of three amylase fractions. Electrophoresis of hog pancreatic amylase has shown a single zone with a mobility similar to γ-globulin; addition of a diluted preparation of this enzyme to rabbit serum increased the γ-globulin fraction only (Berk et al., 1963). This has given support to the view that rabbit serum contains at least two amylase fractions.

Ujihira et al. (1965) have used a saccharogenic method for studying the electrophoretic and chromatographic patterns of distribution of human serum amylase. They have shown that most of the saccharogenic activity was in the γ-globulin fraction both in normal individuals and those suffering from pancreatitis. Chromatography of human serum on Sephadex G75 has shown that the saccharogenic activity could be easily separated from the serum proteins. Similar results have been reported by Wilding (1963), who has separated human serum amylase as a single peak after chromatography on Sephadex G100. The amylases from human saliva, urine, pancreatic juice, and pig pancreas, as well as from sera of patients with pancreatitis or mumps also appeared to have the same molecular weight on the basis of their elution volumes during gel filtration. The effect of serum proteins on amyloclastic techniques for detection of amylase isoenzymes after serum protein electrophoresis has been examined (Wilding, 1965; Searcy et al., 1964). It would appear that certain serum protein fractions, especially albumin, possess non-enzymic amyloclastic activity. Saccharogenic methods are, therefore, mandatory for the study of serum amylase isoenzymes. The reports of amylase activity in human serum protein fractions other than γ-globulin are to be considered doubtful.*

Support for a pancreatic origin of human serum amylase could be found in studies relating to pancreatectomized dogs and clinical studies of patients with pancreatitis (Berk et al., 1965). Joseph et al. (1966) have used a visual staining technique, based on a saccharogenic method, for amylases after agar electrophoresis and have shown that the enzymes from human liver, pancreas and salivary gland have slightly different mobilities. Normal human serum amylase appeared to have the same mobility as the liver enzyme and in patients with acute pancreatitis the serum enzyme had the same mobility as the pancreatic enzyme.

Comparative studies of the electrophoretic behaviour of serum amylase in the sheep, rabbit, cat, guinea pig, horse and human being have shown the presence of a single amylase component which migrated in

---

* The globulin-bound amylase of Wilding et al. (1964) is a very rare phenomenon.

the γ-globulin position for all these animals except the horse (Searcy et al., 1966). Serum from the rat, pig, cow, dog and goat showed two or three forms of serum amylase.

Oger and Bischops (1966) have shown that amylase from human salivary gland and human pancreas had different electrophoretic mobilities and that duodenal juice and serum contained both isoenzymes. Noerby (1965) had previously found differences in the electrophoretic mobilities of the human salivary and pancreatic enzymes and Sick and Nielson (1964) have been able to demonstrate genetic variations in both the salivary and pancreatic amylases of the mouse.

Urinary isoamylases have been separated by cellulose acetate electrophoresis (Aw, 1966). Two components were normally present, one of salivary origin and the other of pancreatic origin. In urine from patients with acute pancreatitis the pancreatic enzyme was increased, that from patients with mumps showed an increase of the salivary gland enzyme.

Using a rooster antiserum to hog pancreatic amylase, McGeachin and Reynolds (1961) have shown that the amylase activities of hog, rat and dog pancreas were inhibited by 72%, 48% and 8% respectively. The same antiserum inhibited only 39% and 18% respectively of the activities of the hog and dog serum enzymes and had no effect on the rat serum enzyme. The amylases from dog, hog and rat liver were inhibited by 16%, 6% and 5% respectively under the same conditions.

7. GLUTAMINASE

Two types of this enzyme have been described, one which catalysed the hydrolysis of glutamine to give glutamate and ammonia and another which was associated with a transamination reaction and required the presence of keto-acids (Roberts, 1960). The former enzyme was activated by phosphate.

Using extracts of mammalian kidneys, Katunama et al. (1966) have shown that the ratio of the glutaminase activity with and without phosphate varied according to the species and they concluded that glutaminase might be a mixture of a phosphate-dependent and a phosphate-independent enzyme. Chromatography of the tissue extracts on calcium phosphate gel allowed separation of two such isoenzymes. The two differed in many respects; the phosphate-dependent entity was heat-labile and was inactivated by heating to 55°C for one minute, whereas the phosphate-independent isoenzyme was inactivated only by heating to over 70°C. The isoenzymes also differed in pH optima, Michaelis constants and sensitivity to activators such as maleate and carbonate. Dietary protein intake appeared to alter the differential activity of the two isoenzymes; a high intake increased the activity of the phosphate-dependent component, whereas that of the other iso-enzyme did not appear to be controlled by diet.

8. MISCELLANEOUS ENZYMES

Kusakabe and Miyake (1966) have shown that iodotyrosine deiodinase from human thyroid could be separated into two components by starch block electrophoresis. One component was associated with the particulate (mitochondrial and microsomal) fraction and the other with the cytoplasm. No differences in properties such as heat lability and co-enzyme specificity could be detected, although Stanbury (1957) had reported that the microsomal enzyme had a higher Michaelis constant and was more sensitive to heat. A relative increase in activity of the soluble fraction enzyme was observed in hyperthyroid glands, and a relative decrease was noted in carcinoma tissue.

Two peaks of ribonuclease have been described after starch block electrophoresis of human serum; one of the peaks was significantly increased in hypothyroidism (Leeper, 1963).

A starch-gel electrophoresis technique for study of ribonuclease iso-enzymes in human tissue extract and serum has been developed (Ressler et al., 1966). The results have indicated that there might have been differences in the isoenzyme pattern from tissue to tissue and that the serum enzyme could be separated into at least three components, one of which was present in most tissue extracts.

Deoxyribonucleases have been identified after disc electrophoresis in polyacrylamide gel (Boyd and Mitchell, 1965). The patterns in larvae, pupae and flies of D. melanogaster, as well as bovine pancreas and bovine spleen have been demonstrated.

Cory and Wold (1965) have prepared crystalline enolase from the muscle of the rainbow trout. The enzyme could be separated into three forms by starch gel electrophoresis.

Suld and Herbut (1965) have purified asparaginases from guinea pig serum and liver. They differed in pH optima and behaviour during chromatography on DEAE-cellulose. With regard to antitumour activity, the serum enzyme was two to three times more potent than the liver enzyme when used against a transplantable mouse lymphosarcoma.

Chytil (1965) has separated isoenzymes of mammalian β-galactosidase using a visual staining technique after starch gel electrophoresis or gel filtration on Sephadex G100. Two components were apparently in beef liver but only one in rat liver. Using chromatography on DEAE cellulose, Furth and Robinson (1965) have found four fractions of β-galactosidase in rat liver lysosomes, both in the lysosomal sap and membrane.

9. ALDOLASE*

The enzyme from rabbit muscle has been found to be thirty to fifty times more active with fructose 1–6 diphosphate as substrate than with fructose 1-phosphate whereas the enzyme from bovine liver used both substrates with the same efficiency (Blostein and Rutter, 1963). These differences most probably represented tissue specificity rather than species specificity.

Using antisera to muscle aldolase prepared in roosters it has been shown that the enzymes obtained from muscle and liver differed immunologically (Blostein and Rutter, 1963). Chromatography of crude animal tissue extracts on cellulose phosphate columns has provided evidence that aldolase with the kinetic characteristics of both enzymes was present in many tissues. If the enzyme from muscle was treated with carboxypeptidase or crude liver extracts, its kinetic characteristics altered so that it behaved like the liver enzyme.

Using starch gel electrophoresis, Anstall *et al.* (1966) have demonstrated multiple forms of aldolase in human, rat and frog tissues. Five isoenzymes have been described in extracts of human brain, three in human heart and two in human kidney, liver and muscle. In rat brain there were four isoenzymes similar in mobility to those of human brain; the rat muscle pattern differed markedly from that of the rat liver, whereas the only difference found between human muscle and human liver was quantitative in nature.

The isoenzyme patterns were distinctly altered when fructose 1-phosphate rather than fructose-1,6-diphosphate was used as substrate.

---

* Aldolase, which is a lyase, is included here for convenience.

CHAPTER V

Physiological Aspects and Metabolic Role

1. LACTATE DEHYDROGENASE

SUBSTRATE inhibition of lactate dehydrogenase by pyruvate is well established and it has been shown that each of the five isoenzymes behaves differently towards increasing pyruvate concentration. Plagemann *et al.* (1960b) have demonstrated that at any fixed pH and temperature the greater the electrophoretic mobility of an LDH isoenzyme, the lower its K_m for pyruvate and the lower the concentration of pyruvate which inhibits the enzyme. Cahn *et al.* (1962) have shown that the enzyme from heart muscle is subject to substrate inhibition by pyruvate at a much lower pyruvate concentration than that from skeletal muscle. Their findings indicate that the extent of substrate inhibition with pyruvate can be correlated with the percentage of M or H subunits present in a particular lactate dehydrogenase.

Pfleiderer and Wachsmuth (1961) were the first to point out that LDH-5 is more predominant in anaerobically metabolizing tissues such as human liver and skeletal muscle, whereas LDH-1 is more prominent in aerobically metabolizing tissues such as human brain and heart. Similar findings related to other species and based on substrate inhibition by pyruvate (Cahn *et al.*, 1962; Lindsay, 1963; Markert, 1963b) have all indicated a possible relationship between substrate inhibition, metabolic role and sub-unit composition of lactate dehydrogenase. Tissues containing a preponderance of H sub-units will allow pyruvate to accumulate and activate the tricarboxylic acid cycle, whereas tissues containing a preponderance of M sub-units will not allow pyruvate to accumulate but will respire anaerobically and thereby create an oxygen debt (Cahn *et al.*, 1962). In heart muscle the steady supply of energy required is maintained by complete oxidation of pyruvate via the tricarboxylic acid cycle but in voluntary muscles bursts of energy are required and glycolysis is maintained by the enzyme operating even in relatively high concentrations of pyruvate or lactate.

Tricarboxylic acid cycle intermediates have been reported as being activators of LDH-5 (Fritz, 1965) and it has been suggested that lactate accumulation in skeletal muscle after vigorous exercise may be due in part to increased $NADH_2$ from glycolysis and in part to activation of LDH-5 by citric acid cycle intermediates. Markert (1963b) had previously suggested that the fundamental role of lactate dehydrogenase is regulation of the $NAD/NADH_2$ ratio. Activation of LDH-5 would result in more NAD being available for other metabolic pathways (Fritz, 1965).

Recently Vesell (1965c) has reported that lactate dehydrogenase of whole human tissue extracts behaved more similarly to increased concentrations of pyruvate and lactate than would be expected from the differences in substrate inhibition of the individual isoenzymes LDH-1 and LDH-5. He has found that the total LDH activity of a tissue extract did not react with pyruvate according to the proportion of M or H sub-units but rather as a mixture of M or H sub-units in equal amounts. Evidence has also been presented which suggested that at 37°C in the human (Vesell, 1965c) and at 39°C in the dog (Vesell and Pool, 1966) LDH-1 and LDH-5 resemble each other in behaviour towards increasing substrate concentrations. Estimation of lactate and pyruvate concentrations in exercised ischaemic canine muscle has shown that even after severe exercise the lactate and pyruvate concentration of the muscle failed to reach the levels required for substrate inhibition (Vesell and Pool, 1966). The possibility that the intracellular concentration of the substrates at the actual locations of the isoenzymes may reach inhibitory levels could not be excluded but was not considered likely (Vesell and Pool, 1966). Vesell (1965c) has reported other evidence purporting to contradict the view that aerobic metabolism is an influencing factor in LDH synthesis. The LDH patterns of erythrocytes, platelets and bovine lens fibres show mainly the faster moving isoenzymes. This does not appear to fit in with the belief that these tissues are not able to take part in aerobic metabolism. Lindy and Rajasalmi (1966) have questioned whether these cells can be compared with ordinary body cells as they are not capable of protein synthesis.

Further studies on the substrate inhibition by pyruvate of lactate dehydrogenase activity in human tissue extracts have not confirmed the results of Vesell (1965c) but rather supported the theory that the *in vitro* properties of the enzyme from crude tissue extracts can be correlated with the sub-unit composition and probable metabolic role (Latner *et al.*, 1966a). Other workers (Stambaugh and Post, 1966b) have also reported similar findings both in relation to crude tissue homogenates at 25°C and 37°C and to purified LDH-1 and LDH-5 at 37°C.

Product inhibition of lactate dehydrogenase has been reported as possibly having more significance in relation to the metabolic role of LDH isoenzymes (Stambaugh and Post, 1966a). The total lactate dehydrogenase activity of tissue homogenates retains the product inhibition characteristics expected from the LDH sub-unit composition at both 25°C and 37°C (Stambaugh and Post, 1966b). It has been suggested that as the concentrations of pyruvate and lactate reported to occur in contracting muscle were not sufficient for significant substrate inhibition of either LDH-1 or LDH-5 but were sufficient for marked differential product inhibition by lactate, the metabolic role of the isoenzymes could be more dependent on the latter. Product inhibition can be related to the metabolic role of the isoenzymes in a manner similar to that for substrate inhibition. Those tissues containing a preponderance of H sub-units will be adapted to aerobic metabolism, since product inhibition by lactate would direct the pyruvate to oxidation by the tricarboxylic acid cycle. Those tissues containing a preponderance of M sub-units which are not as sensitive towards product inhibition would allow lactate to accumulate.

Evidence supporting the theory of a relationship between sub-unit composition and metabolic role has been provided by a number of investigations on a variety of animal species. Wilson *et al.* (1963) have examined the lactate dehydrogenases from the breast muscle of more than thirty species of adult birds. Using the ratio of activity with 0·33 mM pyruvate to activity with 10 mM pyruvate as an index of the relative proportions of M and H sub-units, it has been possible to show that in birds capable of sustained flight, such as the stormy petrel, humming bird and swift, the breast muscle enzyme was two or three times more active at the lower pyruvate concentration, whereas the enzyme activity from the breast muscles of domestic fowl and game birds was approximately the same at both pyruvate concentrations. Birds which undertake periods of long and sustained flight had the type of muscle lactate dehydrogenase which prevented build-up of lactate within the muscle, i.e. the enzyme contained a high proportion of H sub-units; birds which fly only in short bursts had a muscle lactate dehydrogenase containing a high proportion of M sub-units, which means that during flight the consequent build-up in lactate causes an "oxygen debt" with accompanying fatigue.

Studies of the lactate dehydrogenase isoenzyme patterns of various types of muscles of the rabbit, chicken and human being have also yielded results which showed an excellent correlation between the function of a given muscle and the isoenzyme pattern (Dawson *et al.*, 1964; Kaplan, 1964). Lindsay (1963) has reviewed the relationship between the pattern of the heart and skeletal muscles of a variety of animal species, and their respiratory rate, their pyruvate optima and the muscle colour. He has shown that red muscle fibres had low and white fibres high pyruvate optima. Other workers (Van Wijhe *et al.*, 1964) have demonstrated that red muscle fibres contained relatively more of the faster moving isoenzymes than did white muscle fibres. As Dawson *et al.* (1964) have pointed out, certain red muscles, such as soleus, are in a continuous state of contraction in order to maintain posture and thus would be expected to have a lactate dehydrogenase with a preponderance of H sub-units. In general it appears that muscles which contract rhythmically or tonically have a lactate dehydrogenase with a high proportion of H sub-units, whereas those which are exercised only at intervals have a lactate dehydrogenase with a greater proportion of M sub-units (Dawson *et al.*, 1964). Other evidence which supported a relationship between the lactate dehydrogenase isoenzyme pattern and metabolic role has been presented by Salthe (1965) who has found that the patterns of amphibian hearts showed a positive correlation with oxygen availability in the external medium that normally surrounds the skin. Using the ratio of the LDH activities at 0·33 and 10 mM pyruvate as an index, it has been possible to demonstrate that terrestrial amphibia showed more substrate inhibition than did aquatic amphibia which lived in poorly oxygenated waters. Data obtained from the lungless plethorid salamanders has shown that little substrate inhibition could be demonstrated with heart LDH in these species, since the animals have become adapted to living in a state of relative anoxia (Salthe, 1965).

Tissue lactate dehydrogenase isoenzyme patterns show significant alterations during development from foetus to adult (see Chapter VII). These can be related to changes in physiological environment. In the mammalian foetal heart there is a much greater proportion of M sub-units than in adult heart; this may be due to repression of synthesis of the M sub-units due to the lower environmental oxygen tension.

The effect of hormones on lactate dehydrogenase isoenzyme patterns is of interest in so far as it gives examples of hormonal control of sub-unit synthesis. Richterich *et al.* (1963) have shown that the LDH isoenzyme patterns of the myometrium of pregnant and non-pregnant women differed; the former had a greater proportion of the slower moving isoenzymes. A similar shift in isoenzyme distribution had been previously reported in the rat uterus (Allen, J. M., 1961). During pregnancy (Biron, 1964) the percentage of the "muscle" type of lactate dehydrogenase in the uterine muscle of the rat and the rabbit was increased. Goodfriend and Kaplan (1964) have demonstrated that the administration of oestradiol to immature female rats and rabbits resulted in a selective increase in the slower moving isoenzymes of the uterus. Administration of testosterone to male rats produced similar changes in relation to the seminal vesicles. Progesterone and testosterone administration led to an increase in total lactate dehydrogenase activity in the rat uterus without a selective increase in either the slower or faster moving isoenzymes. It has been suggested that oestradiol prepares the uterus for the prolonged contractions during labour by altering the isoenzyme pattern so that it is more conducive to anaerobic respiration.

The findings after administration of oestradiol have indicated that the hormone increased the synthesis of both M and H sub-units in most tissues but that in certain tissues such as the uterus there was preferential synthesis of the M sub-unit (Goodfriend and Kaplan, 1964). This would indicate that hormonal control of selective sub-unit synthesis occurs only in specific "target" tissues.

Intra-muscular injections on alternate days of 40 μg triiodothyronine to rabbits for a total of six injections has raised the basal metabolic rate by 30% and altered the LDH isoenzyme pattern in the liver (Allison et al., 1964). The change involved loss of LDH-5 activity and a decrease of LDH-2, LDH-3 and LDH-4 activities, in other words a repression of synthesis of the M sub-unit. Another example of changes in lactate dehydrogenase isoenzyme pattern produced by hormones is the alteration during thyroxine-induced tadpole metamorphosis (Kim et al., 1966). In tadpole tail and brain, three isoenzymes are normally present; the addition of thyroxine to a concentration of 0·26 nM to the waterbath containing the tadpoles produced a selective decrease in the slowest moving fraction.

Examination of tissue lactate dehydrogenase isoenzymes in goldfish has shown alterations in the liver pattern during adaptation from a cold (4–5°C) to a warm (20–22°C) water environment (Hochachka, 1965). Most tissues showed increases in total LDH activity during cold adaptation but liver alone showed any alteration of the isoenzyme pattern. In goldfish liver there appears to be an increase in synthesis of the H sub-unit during cold adaptation; this may be correlated with the increase in extra-mitochondrial metabolism known to occur in fish liver under these conditions.

The patterns of lactate dehydrogenase found during tissue culture are also apparently examples of regulation of sub-unit synthesis by oxygen tension. The normal LDH isoenzyme pattern of all tissue cultured cells consists of predominantly slower moving isoenzymes (Chapter II). Tissue culture of chick muscle under varying conditions of oxygen tension has shown that elevation prevented the synthesis of large amounts of the M sub-unit, i.e. the cultured tissues did not contain a preponderance of the slower moving isoenzymes (Goodfriend and Kaplan, 1963; Dawson et al., 1964). Similarly with cultures of chick hearts there was an increase in synthesis of H sub-units when the oxygen tension was increased. Cahn (1963; 1964) has studied the effect of oxygen tension and oxidative substrates on the lactate dehydrogenase isoenzyme patterns of cultured chick heart cells. He has confirmed the results of Goodfriend and Kaplan (1963) in that increased oxygen tension enhances synthesis of H sub-units or retards synthesis of M sub-units. With cultured embryo chick heart cells there was increased synthesis of M sub-units of lactate dehydrogenase; citric acid cycle intermediates and coenzyme A repressed this increased synthesis in the same way as raised oxygen tension (Cahn, 1964). Goodfriend *et al.* (1966) have made a detailed study of the control of lactate dehydrogenase synthesis in tissue cultured cells (Fig. 23) and chick embryos.

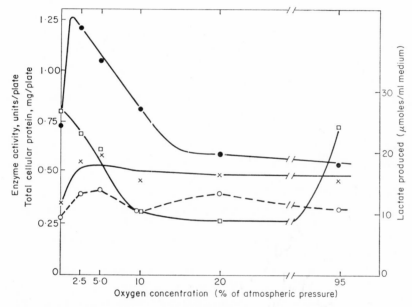

Fig. 23.   Effect of oxygen tension on synthesis of lactate dehydrogenases in tissue cultured monkey heart cells. ●, enzyme activity attributed to M sub-units; ×, enzyme activity attributed to H sub-units; ☐, amount of lactate produced; ○, total cellular protein (reproduced with permission from Goodfriend *et al.*, 1966).

The rate of synthesis of M sub-units appeared to be specifically regulated by oxygen tension. If the oxygen tension fell below 0·1 atmosphere, synthesis of M sub-units was increased. This increased synthesis could be retarded by actinomycin D, puromycin or actidione and also by lowering the temperature from 37° to 4°C. If chelating agents, such as EDTA or 2,2′-bipyridine were added to the system during a period of high oxygen tension, there was an increase in the synthesis of M sub-units which suggested that heavy metal ions regulate the repressive effect of the high oxygen tension. From these results it can be concluded that oxygen tension is not the sole regulator of LDH synthesis (Goodfriend *et al.*, 1966).

An investigation of the response of chick embryo lactate dehydrogenases to variations in the ambient oxygen tension has been reported (Lindy and Rajasalmi, 1966). Eggs were incubated in boxes ventilated with either 40% or 15% O_2 in nitrogen, or air. The variable oxygen tension had no effect on the total enzyme activity of the chick tissues but there was a considerable increase in the proportion of M sub-units in those tissues incubated under hypoxic conditions.

2. MALATE DEHYDROGENASE

The enzymes from the mitochondrial and cytoplasmic fractions of rat liver have been shown to exhibit marked differences in kinetic characteristics (Kaplan, 1961). Oxaloacetate acts as a potent substrate inhibitor of the mitochondrial enzyme at concentrations at which the cytoplasmic enzyme still shows maximal activity. Substrate inhibition by malate is more effective with the cytoplasmic enzyme. This would seem to indicate that the mitochondrial enzyme is better employed in the oxidation of malate, whereas that in the cytoplasm favours the reduction of oxaloacetate. A possible metabolic role for these two isoenzymes has been suggested by Kaplan (1961; 1963). In the cytoplasm, oxaloacetate is reduced to malate which is then oxidized to oxaloacetate in the mitochondria. The oxaloacetate would again be reduced by the cytoplasmic enzyme. The $NADH_2$ produced by the mitochondrial enzyme could be oxidized by respiratory chain enzymes and coupled with formation of ATP (Fig. 24).

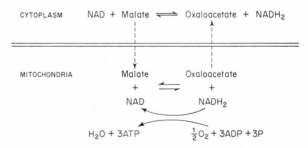

FIG. 24.   Malate dehydrogenase—proposed metabolic role (modified from Kaplan, 1963).

With α-glycerol phosphate dehydrogenase a similar role for the mitochondrial and cytoplasmic enzymes has been described (Bucher and Klingenberg, 1958; Sacktor, 1958). This involves oxidation of

dihydroxyacetone phosphate in the cytoplasm and reduction of
α-glycerophosphate in the mitochondria.

3. ASPARTATE KINASE

Stadtman and colleagues (1961) have studied the inhibition of aspartate
kinase from *E. coli* by L-lysine, L-threonine and DL-homoserine. They
have shown that the degree of inhibition varies with the growth conditions and extraction procedure but that in general the enzyme was
inhibited 30–50% by either lysine or threonine and 10–15% by homoserine. When two amino acids were used as inhibitors simultaneously,
the total inhibition was roughly equal to the sum of that observed for
each independently. These findings have been interpreted as indicating
that extracts of *E. coli* contain three aspartate kinases, one of which is
selectively inhibited by lysine, one by threonine and the other by
homoserine (Stadtman *et al.*, 1961) (Fig. 25). The lysine- and threonine-
sensitive isoenzymes could be selectively precipitated with ammonium

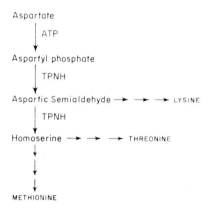

Fig. 25. Aspartokinases—multiple forms each with a different end-product (reproduced
with permission from Stadtman *et al.*, 1961).

sulphate and showed different heat stabilities. Studies of the enzyme
extracted from *E. coli* grown in the presence of 10^{-2} M L-threonine or
L-lysine have shown that the enzyme from the organism grown in the
presence of lysine was no longer sensitive to lysine but that the enzyme
from this organism grown in the presence of threonine had only slightly
reduced threonine sensitivity.

   The three aspartate kinases of *E. coli* are evidently subject to differential regulation by feedback inhibition or repression (Stadtman *et al.*,
1961).

E

4. AMINOTRANSFERASES

Sheid and Roth (1965) have studied the effects of hormones and L-aspartate on the activity and distribution of aspartate aminotransferase activity in rat liver. The administration of six daily injections of 12 mg of cortisone to male rats has been shown to double the aminotransferase activity of liver homogenates. The mitochondrial enzyme showed little change, whereas the specific activity of the cytoplasmic enzyme was increased by some 300%. Administration of cortisol or ACTH did not show such marked effects.

Hormones have also been shown to affect the distribution of alanine aminotransferase in rat liver (Swick et al., 1965). The activity of the mitochondrial enzymes was preferentially increased several fold by prednisolone, while the cytoplasmic and mitochondrial enzymes responded similarly to corticosteroids or alloxan. The activity of the cytoplasmic enzyme was increased by X-irradiation of the rats.

5. CREATINE KINASE

Jacobs et al. (1964) have shown that the creatine kinase from supernatant and mitochondrial fractions of rat tissues could be separated by agar gel electrophoresis. Four isoenzymes A, B, C and D have been demonstrated to exist in extracts of brain, heart and skeletal muscle; the C isoenzyme was present in the mitochondria of all tissues examined. It has been postulated that the presence of relatively high amounts of creatine kinase in mitochondria may indicate a previously undescribed pathway for mitochondrial phosphate metabolism involving a possible role in the transfer of high-energy phosphate between intra- and extra-mitochondrial compartments (Jacobs et al., 1964). Further evidence for a specific role for the mitochondrial isoenzyme has been provided by Bessman and Fonyo (1966) who have shown that it did not react appreciably with intra-mitochondrially bound nucleotides but required addition of adenine nucleotides. Their results did not confirm the role of mitochondrial creatine kinase previously proposed (Jacobs et al., 1964). They suggested that the role of the mitochondrial isoenzyme is the feed-back regulation of respiration in response to muscular activity.

6. HEXOKINASE

Evidence has been presented which indicates a possible relationship between the multiple molecular forms of this enzyme and diabetes. Katzen and Schimke (1965) using starch gel electrophoresis have shown that up to four forms of the enzyme could be detected in extracts of rat liver, brain, kidney, muscle and fat pad. The four isoenzymes have been designated Types I to IV and a correlation between the presence of Type II isoenzyme and insulin sensitivity of the tissue has been noted (Katzen and Schimke, 1965). The Type II hexokinase is decreased in diabetic animals and also in the adipose tissue of normal rats during starvation. McLean *et al.* (1966) have also examined hexokinase patterns in the epididymal fat pad from normal and alloxan-diabetic rats. Isoenzymes of Types I and II have been detected; there was significantly lower total hexokinase activity in the fat pad of diabetic rats but there was a greater loss of hexokinase Type II than Type I. Further observations by Katzen (1966) have shown that the Type II hexokinase of fat pad does not appear to be lowered by diabetes if the enzyme is examined in the presence of mercaptoethanol. In the absence of mercaptoethanol and EDTA the isoenzyme could be separated into two forms by starch gel electrophoresis. Similar results have been obtained with other tissues of diabetic rats. It has been speculated that the insulin effect may be associated with Type II hexokinase *via* a thiol-disulphide interchange reaction between the hormone and this form of the enzyme.

7. HYDROLASES

Starch gel electrophoresis of extracts of tissues of the female rat reproductive tract has shown that the activity of one of the alkaline phosphatase components varied during the oestrus cycle. Three other phosphatases, four acid phosphatases and seven esterases did not appear to be under direct hormonal control (Robboy and Kahn, 1964). One of the alkaline phosphatases of human RA amnion cell cultures was also apparently under hormonal control (Beckman and Regan, 1964). Changes in human serum alkaline phosphatase isoenzyme patterns during pregnancy have been discussed earlier (see Chapter IV).

Hormonal control of esterases has also been indicated in some cases. Allen and Hunter (1960) have demonstrated an association between male sex hormone and certain esterases in the mouse epididymis. An esterase in mouse kidney was also dependent on the male sex hormone (Shaw and Koen, 1963). Changes in the serum esterase patterns have been detected in the pregnant rabbit (Hunter *et al.*, 1964). During the third week of pregnancy there was a progressive loss of certain components, which continued beyond parturition.

8. OTHER ENZYME SYSTEMS

Umbarger and Brown (1958) have shown that two enzymes in *Aerobacter aerogenes* catalyse the conversion of two molecules of pyruvate to α-acetolactate; the α-acetolactate formed could be a precursor of either valine or acetoin (Fig. 26). One form of the enzyme was synthesized in the organism when it was grown at neutral or alkaline pH. This was under specific regulation by valine through feed-back repression. The other form was synthesized only when the organism was grown at acid pH but was not subject to regulation by acetoin.

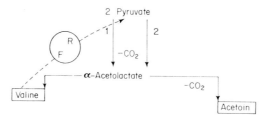

Fig. 26.   Acetolactate metabolism (reproduced with permission from Stadtman, 1963).

A similar phenomenon has been described in relation to the threonine deaminase of *E. coli* (Umbarger and Brown, 1957). One form of the enzyme appeared to be selectively regulated by isoleucine, whereas the other form was not so affected by metabolic end products (Fig. 27).

A similar situation to that found with aspartate kinases has been reported with respect to the enzyme catalysing the condensation of phosphoenol pyruvate and D-erythrose-4-phosphate in *E. coli* (L. C.

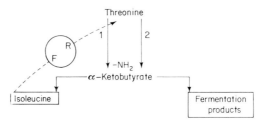

Fig. 27.   Threonine deaminase (reproduced with permission from Stadtman, 1963).

Smith *et al.*, 1962). This is the first step in the biosynthesis of tyrosine and phenylalanine. It has been found that the two aromatic amino acids each control a different form of the enzyme catalysing this condensation (Fig. 28). The two forms could be separated by ammonium sulphate fractionation and heat stability.

The study of aromatic amino-acid biosynthesis has been extended by Doy and Brown (1965) who have separated three isoenzymes of phospho-2-oxo-3-deoxy heptonate aldolase (PODH aldolase) from *E. coli* by ammonium sulphate fractionation. Isoenzyme Ia was inhibited by phenylalanine and isoenzyme Ic by tyrosine. Further investigations of these isoenzymes in *E-coli W* and derived mutants have shown that synthesis of Ia was repressed by phenylalanine, that of Ib by tyrosine and that of Ic by tryptophan (Brown and Doy, 1966). There was also evidence of a certain amount of cross repression, e.g. Ia was also moderately repressed by tryptophan.

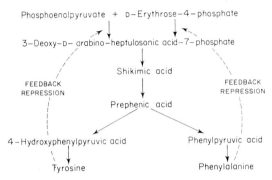

Fig. 28.   Aromatic amino-acid metabolism (reproduced with permission from Stadtman, 1963).

CHAPTER VI

Genetic Studies

THERE IS NOW much information accruing which is related to the genetic control and genetic variation of isoenzyme patterns. The main isoenzyme systems studied have been lactate dehydrogenase, glucose-6-phosphate dehydrogenase, isocitrate dehydrogenase, phosphogluconate dehydrogenase, alkaline phosphatase, acid phosphatase, phosphoglucomutase, adenylate kinase and certain esterases.

This section does not deal with genetic studies in relation to insects, plants and lower life forms since an account of these is given elsewhere (Chapter VII).

1. LACTATE DEHYDROGENASE

If each of the two polypeptide sub-units of this enzyme is controlled by a separate gene, one would expect the possibility of gene mutations and consequent genetic variations of both A(M) and B(H) sub-units resembling those occurring with the haemoglobins.

In the human being, Boyer and co-workers (1963) have described a variant of the B sub-unit demonstrable by starch-gel electrophoresis, in the red cell haemolysates obtained from a 25-year-old Nigerian male; see Fig. 29. Multiple components appeared within the major lactate dehydrogenase zones. Instead of a single band, LDH-1 consisted of five

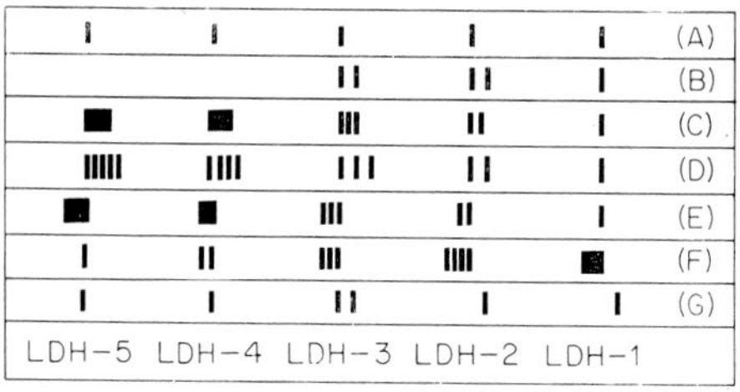

FIG. 29. Lactate dehydrogenase isoenzyme patterns after starch gel electrophoresis. Comparison of the normal (A) and variant (B–G) forms. These variants have been reported as follows: (B) Nance et al. (1963), (C) Kraus and Neely (1964) [Memphis–1]; Davidson et al. (1965), (D) Kraus and Neely (1964) [Memphis–2]; Latner (1964), (E) Kraus and Neely (1964) [Memphis–4], (F) Kraus and Neely (1964) [Memphis–3]; Boyer et al. (1963), (G) Vesell (1965d).

The variants are shown in this diagrammatic form which shows all the possible sub-bands. In practice some of the sub-bands have not been detected; in most cases erythrocytes have been used as the source of the enzyme and the low levels of LDH-4 and LDH-5 prevent detection of sub-bands of these isoenzymes. The precise nature of LDH-4 and LDH-5 in variant (B) has yet to be established.

components, LDH-2 of four, LDH-3 of three and LDH-4 of two. LDH-5 cannot usually be demonstrated in red cell haemolysates. These findings can be explained by assuming that, in addition to the normal sub-unit B, there is a mutant form which can be designated β. Random combination of the sub-units would then give rise to B_4, $B_3\beta_1$, $B_2\beta_2$, $B_1\beta_3$ or β_4 (LDH-1); B_3A_1, $B_2\beta_1A_1$, $B_1\beta_2A_1$ or β_3A_1 (LDH-2); B_2A_2, $B_1\beta_1A_2$ or β_2A_2 (LDH-3); B_1A_3 or β_1A_3 (LDH-4). Had it been possible to demonstrate an LDH-5 this, of course, would have been only one entity, viz. A_4, since no mutant form of sub-unit A was present. The number of components of each isoenzyme thus derived fits in exactly with the actual findings. Another group of workers (Nance *et al.*, 1963) have demonstrated a variant in the red cells which occurred in four individuals from two generations of a Brazilian family. In this case, however, the products of the normal and mutant allele apparently did not associate randomly to form isoenzymes. The mutant haemolysates showed only two bands in each of the components LDH-2 and LDH-3 and one band in LDH-1.

A genetic variant involving sub-unit A has also been described both in the red cell haemolysates and in extracts from a carcinoma of the cervix in a member of a British family; other members of the family also showed similar abnormality in their haemolysates (Latner, 1964). The LDH-1 activity was present as one band, LDH-2 as two bands and LDH-3 as three bands. In the red cell haemolysates LDH-4 and LDH-5 could not be demonstrated, a fact which again is not unusual in haemolysates. The LDH-4 in the carcinoma of the cervix appeared as a broad, blurred band in which it was unfortunately impossible to count the number of components. It has been possible, however, using borate buffer for electrophoresis, to separate out LDH-5 and demonstrate that it consisted of five bands (Latner and Turner, unpublished observations). It seems that the most likely explanation for this variant is the presence of a mutant form of sub-unit A.

Three new easily identified electrophoretic variants and one similar to that described by Boyer *et al.* (1963) have been detected by Kraus and Neely (1964). These four variants were designated Memphis-1, Memphis-2, Memphis-3 and Memphis-4 and detailed pedigrees for three generations were described. The three new variants involved the A sub-unit, as shown in Fig. 29. Kraus and Neely (1964) detected variants in eight individuals amongst a total of 940. This population group consisted of 610 Negroes and 330 Caucasians and the frequency of variation in the enzyme has been calculated as 0·9%. There was no evidence for a linkage with the ABH blood group system or the Fy^a gene, nor for any linkage with haemoglobin type, Rh system, MNS,

Kell or Sutter blood groups. Davidson and his colleagues (1965) have described a variant of the A sub-unit in two out of 1015 unrelated English people. The pattern could be detected in erythrocytes, leucocytes, sperm and tissue cultured fibroblasts. Family studies have shown this variant to exist in a further twenty-nine individuals and the evidence from the pedigrees suggests that the manner of distribution of the abnormal phenotype is under the control of a relatively uncommon autosomal gene. Three examples of a variant involving the A sub-unit have also been observed in a group of 600 Negroes (Vesell, 1965d) one of whom suffered from chronic lymphocytic leukaemia. The pedigree of one of the three variants revealed six affected individuals in four generations and suggested autosomal co-dominant inheritance of the variant phenotype. These variants appear to be identical with the Memphis-1 of Kraus and Neely (1964). Another abnormal lactate dehydrogenase has been reported in a white woman suffering from lymphoblastic sarcoma. In this case LDH-1 and LDH-2 migrated more rapidly towards the anode than normal, while LDH-3 showed normal mobility[1] (Vesell, 1965d). It has been suggested that in this case both *b* alleles were affected and this was a homozygous variant, whereas all the other variants which have been described were heterozygous. Davidson *et al.* (1965) have also found variants involving the A sub-unit in two individuals in a group of 245 Turkish Cypriots and one such variant in twenty-three Ibadan Nigerians. Neel *et al.*, quoted by Vessel (1965b), found no LDH variants in seventy-nine Brazilians. No variants were found by Tashian (Vessel, 1965b) in 284 American Indians, 238 Micronesians and twenty-eight U.S. Caucasians, although he did find one LDH variant in ninety-five U.S. Negroes.

All these population studies have revealed a higher frequency of variants among Negroes and also a greater number of A than B variants. The normal and variant patterns are shown diagrammatically in Fig. 29, and the patterns are classified with references to the authors, who have described them.

Mutations of the enzyme have been detected in a number of animal species. A variant form of the B sub-unit in *Peromyscus maniculatis* was the first genetic variant of lactate dehydrogenase to be described and this was demonstrated by starch gel electrophoresis of extracts of kidney and brain from certain stocks of that animal (Shaw and Barto, 1963). A variant of the A sub-unit has been demonstrated in the laboratory mouse (Costello and Kaplan, 1963). In studies of the polymorphism of lactate dehydrogenase in gelada baboons, three different patterns, including two variants, have been observed in a group of twenty-one

_____

[1] A sub-band of LDH-3 can, however, be detected.

animals (Syner and Goodman, 1966). Variant I could be detected in
one female only and the pattern seemed to indicate a homozygous
variant of the A sub-unit. Variant II was present in seven of the twenty-
one individuals of *Theropithecus gelada* examined. Multiple forms of
LDH-3, LDH-4 and LDH-5 could be detected in extracts of liver,
brain, lens, retina, erythrocytes and plasma. A mutation of the A sub-
unit again seemed to be the most probable explanation; the variant
being heterozygous.

The biosynthesis of lactate dehydrogenase isoenzymes in pigeon testes
has been found to be under the influence of three district gene loci
(Zinkham *et al.*, 1964b). Three phenotypes were found after electrophoresis of homogenates of the testes from racing homer pigeons, wild
park pigeons and Silver King pigeons. Dissociation and recombination
experiments revealed that the results could most easily be explained by
assuming genetic control at three loci A, B and C (Zinkham *et al.*,
1964b). It would appear that all the pigeons so far tested were heterozygous at the C locus. The distribution frequencies of the three types of
pattern provided further evidence for this.

It has already been pointed out (Chapter II) that multiple forms of
lactate dehydrogenase are present in the tissue of the speckled trout,
Salvelinus fontinalis, and the lake trout, *Salvelinus namaycush* (Goldberg,
1965b; 1966). It has been proposed that three non-allelic genetic loci
influence the synthesis of the LDH sub-units and that there should in
fact be fifteen different isoenzymes, if the three sub-units were arranged
in all the possible groups of four, as shown in Fig. 30.

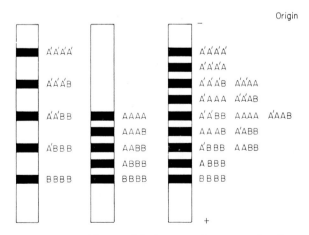

Fig. 30.   Genetic variation of lactate dehydrogenase in trout—sub-unit composition (reproduced with permission from Goldberg, 1966). Hybrid form on the right; it is assumed that
some bands consist of more than one isoenzyme.

E*

These investigations have indicated that nine bands occur only in hybrid species of trout, while in the homozygous species there are five forms of the enzyme. In tissues of the hybrid splake, produced by fertilizing lake trout eggs with speckled trout sperm, nine isoenzymes have been described. The possibility of a common gene locus for LDH-1 has been suggested which would give the observed pattern of nine isoenzymes. Hybridization *in vitro* of lactate dehydrogenase from both homozygous species has also shown the presence of nine bands. At least two extra LDH isoenzymes have been detected in extracts of trout eyes; these could not be found in extracts of other tissues and may also be under genetic control (Goldberg, 1966).

Independent observations by Hochachka (1966) have led to the proposal of a model involving five kinds of sub-unit in these species of fish. Three sub-units designated A, B and C, apparently give the nine isoenzymes previously described. AC hybrids have not been detected *in vivo* but have been produced *in vitro*. Experiments with dissociation and recombination of LDH from fish muscle have shown that random assembly of sub-units in tetramers can give fifteen isoenzymes and it is suggested that the *in vivo* pattern of nine is due to some control mechanism for sub-unit assembly preventing hybridization of certain sub-units. Another two sub-units D and E act independently of the other three sub-units to yield a further five isoenzymes in fish muscle. These isoenzyme patterns are illustrated in Fig. 31.

Genetic variants at the B locus of Merluccius (whiting) have been described, as shown in Fig. 32, one of the mutant alleles being present about as frequently as the normal (Markert and Faulhaber, 1965).

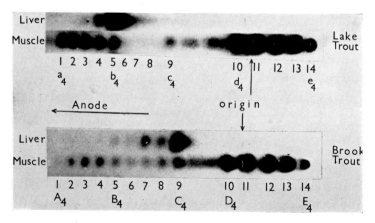

FIG. 31.   Genetic variation of lactate dehydrogenase in trout—presence of five sub-units (reproduced with permission from Hochachka, 1966).

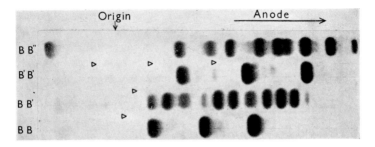

FIG. 32. Genetic variation of lactate dehydrogenase in whiting (reproduced with permission from Markert and Faulhaber, 1965).

Whiting appears to be most similar to mammals in its isoenzyme patterns, dissociation and recombination experiments have shown that the five major isoenzymes are made up of tetramers. In many species of fish additional minor isoenzymes have been detected in extracts of the gonads or eyes; in gonads an additional group of five isoenzymes may be under the control of two genes separate from those in somatic tissues.

2. ISOCITRATE DEHYDROGENASE

Two allelic forms of the enzyme have been described in inbred strains of mice (Henderson, 1965). The variation could be detected only in the enzyme from the supernatant fraction of tissue homogenates. In a heterozygote containing both alleles, three forms of the enzyme have been demonstrated in the ratio of 1:2:1 where the fraction in greatest concentration was the hybrid isoenzyme (see Fig. 17).

3. PHOSPHOGLUCONATE DEHYDROGENASE

Genetically determined forms of phosphogluconate dehydrogenase (PGDH) have been demonstrated by starch-gel electrophoresis of human haemolysates (Fildes and Parr, 1963). The normal phenotype consists of a single zone of activity but in ten out of 150 adults of European descent a second band was detected. This variant was present in about one in twenty-seven of a London population (Parr and Fitch, 1964). A further heterozygous condition in an English family has been described in which a partial quantitative deficiency of PGDH activity was demonstrated (Parr and Fitch, 1964). A similar inherited partial deficiency has been reported in a Negro family (Brewer and Dern, 1964). A simple procedure for differentiation between the normal and variant enzymes has been developed which depends on the relative stabilities of the isoenzymes towards incubation at 37°C for 15 minutes (Parr and Parr, 1965).

A more detailed investigation of PGDH polymorphism has indicated that 95% of unrelated English subjects possess the normal isoenzyme (A) and a total of seven variants made up the other 5% of the normal population (Parr, 1966). Four per cent of these possessed the "common variant" with two major zones and one faint zone of enzyme activity. Approximately 1% of normal individuals possessed a single enzyme band A which was half as active as the normal enzyme. It has been suggested that the genotype for the normal enzyme is $PGD^{\alpha} PGD^{\alpha}$, that for the deficient enzyme $PGD^{\alpha} PGD^{0}$ and that for the common variant $PGD^{\alpha} PGD^{\beta}$ (Parr, 1966). Another variant showing the same isoenzymes bands as the "common variant" but with different intensities, has been designated the "Canning" variant and family studies have indicated that this was homozygous for the PGD^{β} allele (Parr, 1966). The "Canning" variant also existed in the partially deficient state and was apparently heterozygous for the PGD^{β} and PGD^{0} alleles. The so-called "Richmond"

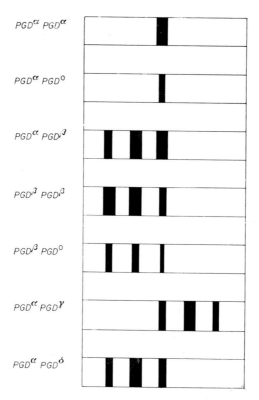

Fig. 33.   Polymorphism of human red cell phosphogluconate dehydrogenase; anode to the right (reproduced with permission from Parr, 1966).

and "Hackney" variants each consisted of three isoenzymes with varying mobilities and they have been provisionally attributed with the
PGD^x PGD^y and PGD^x $PGD^δ$ genotypes. Finally two individuals have
been described in which no PGDH activity could be detected (Parr,
1966). The normal and variant phenotypes are illustrated in Fig. 33.

Population and family studies in the United States and Mexico of
PGDH polymorphism have shown the presence of three phenotypes A,
AB and B (Bowman et al., 1966). These corresponded to the PGD^x PGD^x,
PGD^x $PGD^β$, and $PGD^β$ $PGD^β$ genotypes already described and these
data on American populations confirmed that these isoenzyme phenotypes are under the control of two alleles at a single autosomal locus.

Three phenotypes have been shown to be present in rat erythrocytes.
They consisted of either a single fast or slow moving isoenzyme or a
triplet consisting of both the fast and slow moving isoenzymes, plus one
of intermediate mobility (Parr, 1966).

4. GLUCOSE-6-PHOSPHATE DEHYDROGENASE

The well known relationship between certain types of haemolytic anaemia
and a genetically determined deficiency in erythrocyte glucose-6-phos-
phate dehydrogenase has stimulated a number of investigations of this
enzyme.

Using starch gel electrophoresis, qualitative variants of erythrocyte
and leucocyte glucose-6-phosphate dehydrogenase have been demonstrated in healthy American Negroes but not in Caucasians (Boyer et
al., 1962; Kirkman, 1962; Davidson et al., 1963). The differences in
electrophoretic mobility appeared to be sex-linked. Healthy Negro
males possessed either a fast moving (A) or slow moving (B) variety of
the enzyme and Negro females possessed either the A or B enzymes
or both; all Caucasians possessed the B variant (Boyer et al., 1962;
Kirkman and Hendrickson, 1963).

A rare slow moving variant "Baltimore", associated with normal
erythrocyte enzyme activity, has been detected in an American Negro
population (Boyer et al., 1962). A further very slow moving variant
"Ibadan" has been found in a Nigerian population (Porter et al., 1964).
This variant was also associated with normal levels of the erythrocyte
enzyme. An apparently healthy Italian family with a deficiency of
erythrocyte G6PDH activity has been shown to possess an erythrocyte
enzyme which moved faster than the normal enzyme during starch-gel
electrophoresis. Examination of purified preparations of the normal and
deficient enzymes has shown differences in their catalytic properties, as
well as electrophoretic mobility (Marks et al., 1961). With rare excep-
tions, Negroes with the erythrocyte enzyme deficiency possessed the A

electrophoretic variant. Caucasians with the deficiency possessed the B variant (Boyer et al., 1962; Kirkman and Hendrickson, 1963).

A "Seattle" variant of the enzyme has been described in a healthy male of Welsh-Scottish ancestry. This variant had a lower electrophoretic mobility than the normal enzyme and slightly abnormal kinetic characteristics (Kirkman et al., 1963). Two slow moving variants "Austin 1" and "Austin 2" have been detected in an American Negro.

Kirkman et al. (1965) have reviewed abnormal glucose-6-phosphate dehydrogenases and have indicated the disparities which exist between levels of the erythrocyte enzyme and the severity of the clinical manifestations. They compared the enzyme characteristics and the clinical status in relation to the "Chicago" variant (Kirkman et al., 1964a), the "Oklahoma" variant (Kirkman and Riley, 1961) and the "Mediterranean" variant (Kirkman et al., 1964b).

Kirkman et al. (1964c) have shown that the erythrocyte G6PDH from some Chinese individuals with enzyme deficiency had an electrophoretic mobility in starch gel slightly faster than normal but not as fast as the A variant present in Negroes with the enzyme deficiency. A number of G6PDH electrophoretic variants have been reported in a Chinese population, all the variants being related to some form of haemolytic disorder (Wong et al., 1965).

Evidence for four types of erythrocyte G6PDH has been obtained as a result of studies involving a number of physicochemical parameters (Pinto et al., 1966). Three of these variants occurred in drug-sensitive non-anaemic Negroes; Type I was similar to the normal Negro enzyme with electrophoretic mobility of the usual B type, Type II and III both had electrophoretic mobility of the A type but could be differentiated by other parameters, such as substrate specificity, K_m with 2-deoxyglucose phosphate and activation energy. Type IV enzyme was detected in Caucasians with congenital non-spherocytic haemolytic anaemia. It had electrophoretic mobility of the A type population (Long et al., 1965). Although the erythrocyte enzyme activity was lower than normal, those individuals possessing this variant were not anaemic. Examination of the G6PDH in Iraqi Jewish mutants has demonstrated a so-called "Tel-Hashomer" variant with abnormal kinetic properties as well as altered electrophoretic mobility (Ramot and Brok, 1964; Ramot et al., 1964). The normal and variant enzymes were, however, immunologically identical. An example of the "Baltimore" variant has been described in Papuans but no variants have yet been reported in American Caucasians, Thailanders, Japanese or Marshall Islanders (Porter et al., 1964).

In a Sardinian population, Porter et al. (1964) have found electro-

phoretic variants termed "Sardinia 1" and "Sardinia 2" which involved
an isoenzyme additional to the B variant. They have also detected a
variant similar to the Baltimore variant which was termed "Sardinia/
Baltimore". Similar variants have not, however, been found by Vecchio
et al. (1966) in another Sardinian population.

An unusual "Chicago 1" glucose-6-phosphate dehydrogenase has
been reported in some cases of congenital haemolytic disease; this variant had normal electrophoretic mobility and kinetic characteristics but
was extremely unstable (Kirkman *et al.*, 1964a). Some Sephardic Jews
and Caucasians have been shown to possess an abnormal enzyme with
altered kinetic properties but normal electrophoretic mobility (Kirkman
et al., 1964b), and could be distinguished from the other three types by
K_m values and thermal stability. Nance and Uchida (1964) have found
a genetically determined variant "Madison" with altered electrophoretic
mobility in monozygote twins with XO karyotypes (Turner's syndrome). This variant has also been detected in several relatives of these
twins but not in ten other individuals with Turner's syndrome or their
parents. Thirteen different phenotypes of G6PDH have been described
by Porter *et al.* (1964) on the basis of electrophoretic and enzymatic
variation. Since the kinetic constants of the normal and variant enzymes did not differ in many cases, it would appear that the molecular
differences which altered electrophoretic mobility existed at sites other
than the *catalytic* sites (Long *et al.*, 1965).

Porter and his colleagues (1964) have suggested that the loci controlling the quantitative and qualitative variations may be located
close to each other on the chromosome and they have also found no
evidence that *Plasmodium falciparum* is involved in the polymorphism
of this enzyme.

The electrophoretic mobility of the G6PDH isoenzymes in skin fibro-
blast cultures from normal and enzyme deficient subjects has shown the
same differences as in erythrocytes and leukocytes and there appears to
be a close correlation between erythrocyte and skin enzyme activities
(Nitowsky *et al.*, 1965). Studies of the G6PDH in human females heterozygous for the normal A and B variants have shown that there are two
distinct populations of cells with one clone producing the A variant and
the other the B variant (Davidson *et al.*, 1963). Further investigations
of the distribution of A and B variants in epithelia of heterozygous
Negro females have indicated that the average patch size (adjacent cells
of one type) was less than 0.3 cm^2 (Linder and Gartler, 1965a). These
results have led to estimates of cell population size at the time of X-
chromosome inactivation to be from 7×10^4 to over 2×10^5 cells. This
type of mosaicism has been utilized as a cell marker in the study of

leiomyomas (Linder and Gartler, 1965b). These workers have found that all but one sample of normal myometrium possessed both the A and B electrophoretic variants in equal or nearly equal amounts, whereas all leiomyomas possessed only a single A or B variant, indicating that all these tumours arise from single cells.

The X-linkage of G6PDH in human beings has been confirmed by the family studies of Boyer *et al.* (1962) and Kirkman and Hendrickson (1963). Further confirmation of the X-linkage of mammalian G6PDH polymorphism has been obtained from animal experiments. Trujillo and co-workers (1965) have shown that the erythrocytes of female mules and hinnies contained both horse and donkey enzymes; those from male mules with X-chromosomes from their horse mothers contained only the horse enzyme, whereas those from male hinnies with donkey X-chromosomes contained the donkey enzyme. X-linkage of erythrocyte G6PDH in two species of wild hares, *L. europeaus* and *L. timidus* has been confirmed using reciprocal hybrids (Ohno *et al.*, 1965). Each male hybrid possessed a single G6PDH isoenzyme with the electrophoretic mobility of the enzyme from his mother while both parental enzymes could be detected in female hybrids.

Two forms of G6PDH have been reported in tissues of the deermouse, *Peromyscus maniculatus* (Shaw and Barto, 1965). In all tissues except erythrocytes, two zones of the enzyme have been demonstrated after starch gel electrophoresis; in erythrocytes only the faster moving or A

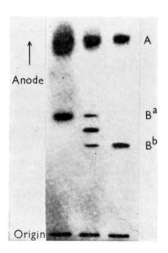

Fig. 34.   Polymorphism of phosphogluconate dehydrogenase in *Peromyscus* (reproduced with permission from Shaw and Barto, 1965).

variety could be detected. In three out of the four deermouse families examined, polymorphism of the B enzyme was apparent. Three pheno-types have been described; Ba consisting of a single zone, Bb consisting of a slower moving single zone, and Bab consisting of three isoenzymes, two with the mobility of the single parental types and a third more intense zone with intermediate mobility (Fig. 34). Genetic evidence has been presented which supports the hypothesis that the enzyme is under the control of a gene with two autosomal alleles. Shaw (1966) has recently reported that this B enzyme with its autosomally controlled polymorphism was equally active with glucose or galactose-6-phos-phates as substrate, whereas the erythrocytic A enzyme was specific for glucose-6-phosphate. Genetic variants of the A isoenzyme in *Peromyscus* erythrocytes have not been detected (Shaw, 1966).

5. ADENYLATE KINASE

Three phenotypes of adenylate kinase have been described in human erythrocytes, heart and skeletal muscle (Fildes and Harris, 1966). They have been designated AK1, AK2–1 and AK2, and are illustrated in Fig. 35. About 10% of the normal population showed the AK2–1

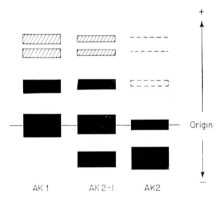

FIG. 35. Genetic variation of human erythrocyte adenylate kinase (reproduced with permission from Fildes and Harris, 1966).

phenotype and only one female out of a mixed population of 960 showed the AK2 phenotype. Family studies on the offspring of fifty-four matings where one parent was AK1 and the other AK2–1 have shown that of 136 children tested seventy-two were AK1 and sixty-four AK2–1 (Fildes and Harris, 1966). These results and those on 141 children from sixty-two AK1–AK1 matings where all the children were of the AK1 pheno-type have indicated that the isoenzymes are controlled by a pair of autosomal allelic genes AK^1 and AK^2, AK1 and AK2 individuals being homozygotes (AK^1AK^1 and AK^2AK^2 respectively) and AK2–1 individuals being heterozygotes (AK^2AK^1).

6. Phosphoglucomutase

Polymorphism of this enzyme has been demonstrated in human red cell haemolysates (Spencer *et al.*, 1964). Several distinct isoenzymes occurred in each individual examined and all the zones present in haemolysates could be detected after starch-gel electrophoresis of extracts of leucocytes, liver, kidney, heart muscle, uterine muscle, brain, skin and placenta. The placental enzyme has been shown to be the same type as that present in cord blood but could differ from that of the maternal circulation. Three phenotypes designated 1, 2 and 2–1 comprising variations in seven individual isoenzymes have been described in a British population (Spencer *et al.*, 1964). These phenotypes are illustrated in Fig. 36 and genetic evidence suggests that the variations are due to a

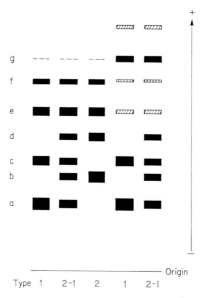

Fig. 36. Polymorphism of human erythrocyte phosphoglucomutase (reproduced with permission from Hopkinson and Harris, 1965).

pair of autosomal alleles, PGM^1 and PGM^2. The slower moving isoenzymes during starch gel electrophoresis appear to be controlled by the PGM^1 gene and the faster moving by the PGM^2 gene.

Five further phenotypes have been detected after extended population studies (Hopkinson and Harris, 1965). Some rare variants of the 1 and 2–1 phenotypes appear to be dependent on genes at two distinct and not closely linked loci, each locus specifically affecting different components of each phenotype. Other uncommon alleles PGM^3, PGM^4 and PGM^5 provide the other abnormal patterns which have been described and are shown in Fig. 37. Another independent investigation of PGM

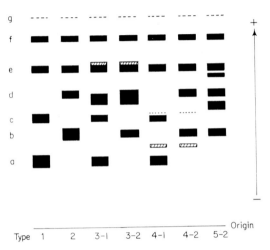

Fig. 37. Rare phenotypes of human erythrocyte phosphoglucomutase (reproduced with permission from Hopkinson and Harris, 1965).

polymorphism led to the recognition of an abnormal phenotype in a Negro population which appeared to be similar to one of the variants previously reported (Luang Eng, 1966).

7. ESTERASES

A. ERYTHROCYTES

The non-specific esterases of human erythrocytes have been examined by starch-gel electrophoresis and shown to exist in multiple forms (Tashian, 1961). Nine zones have been described, eight of which are most active with the shorter acyl chain esters as substrates. Further studies have resulted in these esterases being split into a number of groups designated A_1, A_2, B, C and D (Shaw et al., 1962). An altered form of the D esterase has been described in the erythrocytes of a mentally deficient child (a 4-year-old male mongoloid, with forty-seven chromosomes and a trisomy-21 karyotype). The occurrence of this atypical D esterase has been shown to be under genetic control. Both

the normal and atypical enzyme have been identified as carbonic an-
hydrases on the basis of substrate specificity and sensitivity to inhibitors
(Shaw *et al.*, 1962). A further variant of the D esterase or carbonic
anhydrase has been described in a Micronesian population (Tashian
et al., 1963); this atypical enzyme has a slower mobility than the other D
esterases. Tashian and Shaw (1962) have also revealed the presence of a
variant in the A esterase group; the atypical A esterase having a greater
mobility than the normal enzyme during starch-gel electrophoresis.

Thirteen zones of esterase have been visualized after starch-gel
electrophoresis of erythrocytes from the woodland deermouse, *Peromyscus
maniculatus gracilis*. Four different phenotypes, as shown in Fig. 38 have

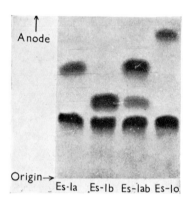

FIG. 38.   Erythrocyte esterases in *Peromyscus*—genetic variations (reproduced with permission from Randerson, 1965).

been described. These involved two esterase components; they appeared
to be under the control of one autosomal locus with three alleles $Es1^a$,
$Es1^b$ and $Es1^0$ (Randerson, 1965). Studies of the 855 offspring from the
ten possible matings have indicated that the $Es1^a$ and $Es1^b$ alleles were
co-dominant, while $Es1^0$ was recessive.

Eleven esterase components have been described after starch-gel
electrophoresis of haemolysates from the house mouse *Mus musculus*
(Pelzer, 1965). Two electrophoretic forms of acetyl esterase Ee1a and
Ee1b, could be detected, as shown in Fig. 39, and the mice could also be
classified by the presence or absence of the propionylesterase Ee2a. The
two Ee1 esterases were shown to be controlled by two co-dominant alleles
$Ee1^a$ and $Ee1^b$ and heterozygotes possessed both the fast and slow moving
forms of the enzyme. The Ee2a esterase was shown to be controlled by
a single autosomal dominant gene $Ee2^a$; the absence of this enzyme
being determined by the recessive allele $Ee2^0$ (Pelzer, 1965).

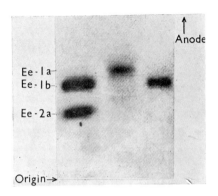

FIG. 39.   Genetic variation of erythrocyte esterases in *Mus musculus* (reproduced with permission from Pelzer, 1965).

Studies of the erythrocyte esterase patterns of domestic rabbits (*Oryctolagus cuniculus*) have resulted in the demonstration of three separate zones of enzyme activity after starch-gel electrophoresis (Grunder *et al.*, 1965). Although no variations in the single component of Zones I and II have been detected, three phenotypes involving five individual esterases have been described for Zone III. These have been designated A, B and AB (Fig. 40). Two isoenzymes were believed to be under the

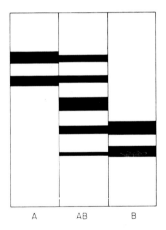

FIG. 40.   Genetic variation of erythrocyte esterases in rabbits (reproduced with permission from Grunder *et al.*, 1965).

control of a pair of autosomal alleles Es^A or Es^B; the heterozygote has been shown to contain five esterase components, two from each parent and a hybrid isoenzyme (Grunder *et al.*, 1965).

B. SERUM ESTERASES

Genetically determined differences in the esterase patterns visualized after starch gel electrophoresis of mouse serum have been reported by Popp and Popp (1962). Three types of pattern could be recognized containing one, two and three components respectively; the three-component pattern resulted from a cross between two mice, one possessing the one-band type and the other the two-band. Petras (1963) has described serum esterase patterns in *Mus musculus*. The fastest migrating of the eleven serum esterases demonstrated was found to be under the control of a pair of incompletely dominant alleles $Es2^a$ and $Es2^b$. Mice with the genotype $Es2^b/Es2^b$ showed a highly active component, those with the genotype $Es2^a/Es2^b$ showed a partially active component, whereas in those with the genotype $Es2^a/Es2^a$ this esterase appeared inactive.

A comparative study of the carboxylic esterases in the sera of the horse, the donkey and their hybrids has shown that sera from the horse, mule and hinny each contain two isoenzymes which hydrolyse α-naphthyl acetate, whereas donkey serum contains only the slower moving of these two esterases (Kaminski and Gajos, 1964). The donkey, therefore, lacks a serum esterase which is present in hybrid species with the horse, whether the donkey is the father or the mother.

The mode of inheritance of plasma esterases in horses has been investigated in relation to other genetic markers such as transferrins, albumins and pre-albumins (Gahne, 1966). Six different esterase patterns have been detected after starch gel electrophoresis of horse plasma at pH 8·5. The phenotype F(fast) was characterized by a fast moving zone with a fainter zone with slightly less mobility. Phenotype I(intermediate) and S(slow) showed similar two-band patterns with different mobilities of the major zone. Phenotypes FI and FS showed esterase patterns corresponding to mixtures of the parental types. The sixth phenotype was characterized by a lack of esterase activity. The data obtained from studies of pedigrees have shown that the esterase phenotypes are controlled by four autosomal alleles of which Es^F, Es^I and Es^S are co-dominant and Es^0 is recessive. This would mean in fact that ten genotypes Es^F/Es^F, Es^I/Es^I, Es^SEs^S, Es^FEs^I, Es^FEs^S, Es^IEs^S, Es^0Es^0, Es^FEs^0, Es^IEs^0 and Es^SEs^0 should exist, although the Es^0 allele is uncommon.

C. HUMAN SERUM CHOLINESTERASE

Investigations which led to recognition of forms of this enzyme under genetic control arose from the finding that certain patients with low serum cholinesterase levels were subject to prolonged apnoea following administration of the muscle relaxant succinylcholine (Bourne *et al.*, 1952; Evans *et al.*, 1952). Succinylcholine is hydrolysed by cholinesterase so that low serum levels of this enzyme cause the period of anaesthesia to be prolonged (Lehmann and Liddell, 1964). A comparison of the action of serum cholinesterase on certain ester substrates, including acetylcholine, has indicated that the rates of hydrolysis of these are always slower with sera having low serum cholinesterase levels than those with normal levels (Davies *et al.*, 1960).

It has now been established that the low enzyme activity was not due to a low concentration of the normal but rather to an abnormal variant, which was less effective in hydrolysing the choline esters; in other words, it had a higher Michaelis constant (Kalow and Genest, 1957; Kalow and Staron, 1957; Kalow and Davies, 1958; Davies *et al.*, 1960; Bamford and Harris, 1964).

The low serum cholinesterase activity appeared to be genetically determined (Forbat *et al.*, 1953; Lehmann and Ryan, 1956) and it has been suggested that a recessive gene was responsible. Further investigations have shown slightly lower values than normal in the heterozygotes (Allott and Thompson, 1956; Lehmann *et al.*, 1958; Kaufmann *et al.*, 1960).

The abnormal enzyme or isoenzyme was particularly ineffective with succinylcholine as substrate and was more resistant than normal to most of the accepted cholinesterase inhibitors (Kalow and Davies, 1958). The local anaesthetic dibucaine has been used as a differential inhibitor of the normal and variant cholinesterases (Kalow and Genest, 1957). The difference between the two cholinesterases was most marked at a concentration of 10^{-5} M dibucaine when the normal enzyme was inhibited by about 80% and the variant about 20% (Fig. 41). This percentage inhibition is commonly referred to as the dibucaine number, which is independent of the total serum enzyme level. In general, a value of this number greater than seventy-one indicates normal esterase, one less than thirty indicates the atypical variant. Intermediate numbers suggest a mixture of normal and atypical isoenzymes (Kalow, 1959). Differential responses of human serum cholinesterase to an inhibitor present in potato peel extracts and later to the glycoalkaloids solanine and solanidine have been described (Harris and Whittaker, 1959; 1962b). The range of substances which differentially inhibit cholinesterases is extensive and include eserine (phystostigmine) and numerous synthetic alkaloids (Kalow, 1959; Kalow and Davies, 1958). Organophosphorus compounds, such as tetraethyl pyrophosphate (TEPP) and di-isopropyl fluorophosphate (DFP) inhibit the normal and atypical esterases to the same extent (Gilman and Koelle, 1949; Kalow and Davies, 1959).

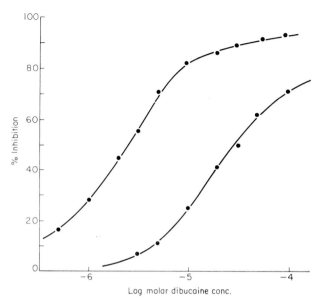

Fig. 41.   Effect of dibucaine on normal (upper curve) and a typical human serum pseudo-cholinesterase (Lehmann and Liddell). "Progress in Medical Genetics" Vol. III, 1964, by permission of Grune and Stratton, Inc.

Family studies have shown that the inheritance of the atypical enzyme is under the control of two co-dominant allelic genes (Kalow and Staron, 1957; Harris *et al.*, 1960; Bush, 1961; Lehmann and Liddell, 1964). Although penetrance of the two alleles appears to be complete, the genes seem to vary in expression in that the dibucaine numbers found in heterozygotes have varied from 48–69 and the corresponding numbers with the inhibitor R02–0683 from 50–80 (Lehmann and Liddell, 1964). There did not appear to be any difference in the gene frequencies in different populations (Kalow and Gunn, 1958; Kattamis *et al.*, 1962; Lehmann and Liddell, 1964).

Further studies of inhibitors of serum pseudocholinesterase have shown that sodium fluoride is also a differential inhibitor (Harris and Whittaker, 1961). With 5×10^{-5} M sodium fluoride, so-called fluoride, numbers of the enzymes could be obtained. In the majority of individuals tested, the enzymes appeared normal or atypical using both dibucaine or fluoride as inhibitors but in some families certain individuals had lower dibucaine or fluoride numbers than expected (Harris and Whittaker, 1961). In other words, there appeared to be a variant with a slightly decreased sensitivity to dibucaine but with a markedly decreased sensitivity to fluoride. A number of geneological studies have shown that the use of dibucaine and fluoride permitted the recognition of six phenotypes (Harris and Whittaker, 1961; 1962a; Lehmann *et al.*, 1963; Liddell *et al.*, 1963; Lehmann and Liddell, 1964). A diagrammatic representation of these serum pseudo-cholinesterase types is shown in Fig. 42.

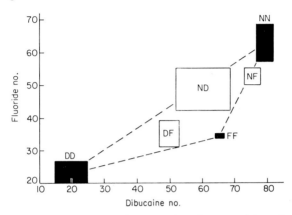

FIG. 42. Fluoride and dibucaine numbers allowing the recognition of six phenotypes (reproduced with permission from Lehmann and Liddell, 1964).

Certain families have been found with anomalous dibucaine values (Kalow and Staron, 1957; Harris *et al.*, 1960; Bush, 1961; Lehmann and Liddell, 1964) and it is possible that these findings may have been due to the presence of a "silent" gene which could determine either complete absence of the enzyme protein or synthesis of a non-enzymatically active protein (Lehmann and Liddell, 1964).

No difference in the mobilities of the normal and atypical enzyme could be detected after starch-gel electrophoresis (Kalow, 1959), although a slight difference in the migration rates of the normal and atypical enzyme could be found if the inhibitor decamethonium was added to the starch gel before electrophoresis (Kalow, 1959). Using a discontinuous buffer system, the major zone of serum cholinesterase has been visualized as a doublet (Dubbs *et al.*, 1960; Harris *et al.*, 1962; Latner, 1962). In patients with apnoea following administration of succinylcholine one of the components of the doublet was apparently not inhibited by eserine and was thus an atypical cholinesterase (Latner, 1962). Using two-dimensional starch-gel electrophoresis, Harris and co-workers (1962; 1963) have been able to demonstrate multiplicity of zones of serum cholinesterase, as shown in Fig. 43. In occasional sera an extra component could be observed which has been designated C5. The incidence of the C5 component has been some 5% of unrelated

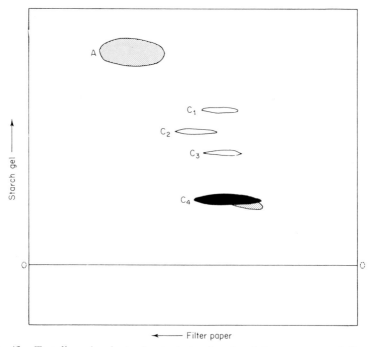

Fɪɢ. 43.   Two-dimensional starch-gel electrophoresis of human serum cholinesterase; A, albumin; C_1, C_2, C_3, C_4, cholinesterases (reproduced with permission from Harris *et al.*, 1962).

British individuals. An additional esterase component was found in cord blood and two components, S_1 and S_2, have been found in certain sera which have been stored in the frozen state (Harris *et al.*, 1963). The component C5 could also be demonstrated after one dimensional starch gel electrophoresis when the normal major component C4 appeared as a doublet with C5. Individuals whose serum possessed the C5 component did not, however, have abnormal dibucaine or fluoride numbers, although the mean level of cholinesterase was about 30% higher in those with C5 components (Harris *et al.*, 1963). Further studies (Robson and Harris, 1966) have shown no relationship between the incidence of the C5 esterase and genetic markers such as haptoglobins, transferrins, erythrocyte acid phosphatases, phosphogluconate dehydrogenase, phosphoglucomutase or adenylate kinase. Family studies have indicated that the presence of this esterase seemed to be determined by an allele in either the heterozygous or homozygous state, although few heterozygotes could be demonstrated by the electrophoretic techniques employed. Two new uncommon patterns of serum cholinesterase after starch gel electrophoresis have recently been described (Van Ros and

Druet, 1966). They involved additional components C_6, C_{7a} and C_{7b} and could be demonstrated by both one- and two-dimensional electrophoresis.

Using column chromatography with gradient elution on DEAE cellulose at pH 9·0, Liddell and co-workers (1962a) have been able to separate the serum enzyme obtained from a heterozygote into two fractions with dibucaine numbers characteristic of the normal and atypical enzymes. Similar separation has also been achieved using paper electrophoresis at pH 9·7. Separation of enzymes with different fluoride numbers has also been obtained (Liddell et al., 1962b).

8. ALKALINE PHOSPHATASE

A second slower moving zone of normal serum alkaline phosphatase was first described by Hodson et al. (1962). It has been suggested that this zone is under genetic control and is present in 60% of normal individuals (Cunningham and Rimer, 1963). In a study of eighty-nine monozygotic and 111 dizygotic twin pairs, Arfors et al. (1963a; 1963b) classified individual samples into Types 1 and 2 on the basis of the absence or presence of the second phosphatase zone. The different serum phosphatase patterns have been found to be strongly associated with the ABO blood group distribution; most individuals with both zones of serum phosphatase being of group O (Arfors et al., 1963a). Further studies have indicated an association between these phosphatase types and the Lewis blood groups. Population studies have also indicated that all individuals with the second serum phosphatase component have been ABH secretors (Beckman, 1964). Schreffler (1965) has confirmed this association between the presence of the second slower-moving band and blood groups and has come to the conclusion that the variation was not an "all or none" phenomenon. The second serum phosphatase could not be detected in individuals who were of blood group A or who were non-secretors of ABH blood group substances. The previously suggested association with the Lewis blood group system has been rejected (Beckman, 1964; Schreffler, 1965). Bamford et al. (1965) have also shown that normal sera contain varying proportions of the second phosphatase. As well as confirming the association with blood groups and secretor status, these workers have found that those individuals with two zones of serum alkaline phosphatase have significantly higher total serum enzyme activities. The amount of the second alkaline phosphatase present in any individual's serum could be influenced by diet (Langman et al., 1966). After a fatty meal, the level of this serum phosphatase was increased and the greatest increase was found in individuals who were O or B secretors.

Polymorphism of the alkaline phosphatase from human placentae has been described (Boyer, 1961; Robson and Harris, 1965). Using starch-gel electrophoresis in two different discontinuous buffer systems, it has been possible (Robson and Harris, 1965) to classify the phosphatase patterns into six types, F, FI, I, SI, FS and S where F, I and S represented fast, intermediate and slow moving isoenzymes respectively. Electrophoresis at pH 8·6 did not allow Types F, FI and I to be differentiated and electrophoresis at pH 6·0 did not allow differentiation of Types I, SI and S. For exact identification of a particular placental phosphatase phenotype, electrophoresis had to be carried out at both pH levels. The patterns obtained are shown in Fig. 44. The six phenotypes appeared to be controlled by three autosomal allelic genes *Pl*, *Pl* and *Pl*. The placental phenotype appeared to be determined

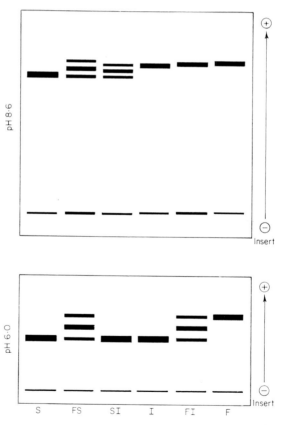

FIG. 44. Polymorphism of human placental alkaline phosphatases as shown by starch-gel electrophoresis (reproduced with permission from Robson and Harris, 1965).

by the foetal genotype. Evidence for this has been obtained by examining the phenotypes of 380 placentae from 190 dizygotic twin pairs; in seventy-eight pairs the phosphatase patterns differed, excluding the possibility that the maternal genotype determined the placental phenotype (Robson and Harris, 1965).

Genetic variation of serum alkaline phosphatase in cattle has been described (Gahne, 1963). Two different types of pattern could be detected, depending on the presence or absence of a fast moving component A. The serum phosphatase patterns appeared to be under the control of two alleles, one giving rise to a pattern including the A component and the other to a pattern without this component. A similar variation to that found in cattle has been reported in sheep (Rendel and Stormont, 1964), in which a so-called B phosphatase was associated with the presence of soluble blood-group substance O in adult females. Further genetic observations have shown a correlation between the presence of the B phosphatase and blood group O and absence of B phosphatase and blood groups R and i (Rasmusen, 1965).

In blood plasma of the fowl, a fast or slow moving zone of activity can be detected after starch-gel electrophoresis (Law and Munro, 1965). Studies of pedigrees of two inbred lines have shown that the expression of the faster moving phosphatase was controlled by a single autosomal co-dominant gene which was allelic to a recessive gene which controlled expression of the slower moving phosphatase. In the heterozygote only the faster moving component was present.

9. Acid phosphatase

Polymorphism of acid phosphatase from human erythrocytes has been demonstrated (Hopkinson et al., 1963; 1964). Five types of two or three band patterns could be recognized and these were designated A, BA, B, CA and CB, as shown in Fig. 45. Patterns similar to those of the naturally occurring type BA could be demonstrated after starch-gel electrophoresis of mixtures of the A and B types (Hopkinson et al., 1964). Population studies have shown that these variants could be under the control of three allelic autosomal genes P^A, P^B and P^C. This would result in six genotypes $P^A P^A$, $P^A P^B$, $P^A P^C$, $P^B P^B$, $P^B P^C$ and $P^C P^C$ (Hopkinson et al., 1963; 1964). The data obtained, however, indicated that the sixth phenotype, C, would be considerably less frequent than the others. It did not occur in this series; it has, however, been demonstrated elsewhere (Lai et al., 1964). Starch-gel electrophoresis in formate buffer pH 5·0 has also been used to produce more distinct erythrocyte acid phosphatase patterns (Giblett and Scott, 1965). Studies of the frequency of the phosphoglucomutase and acid phosphatase variants in Eskimos and Indians have confirmed earlier suggestions (Giblett and Scott, 1965) that the P^C gene appears to be a Caucasian gene (Scott *et al.*, 1966). It has not been found in Oriental populations or in

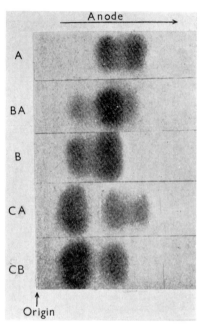

Fig. 45.   Genetic variation of human erythrocyte acid phosphatases (reproduced with permission from Hopkinson *et al.*,
1963).

Eskimos or Indians and only rarely in Negroes. Scott (1966) has purified the $P^A P^A$ and $P^B P^B$ isoenzymes from erythrocytes and studied their kinetic characteristics. The kinetic properties of the two isoenzymes have been shown not to be significantly different.

10. Amylase

Agar-gel electrophoresis of pancreatic extracts and saliva from *Mus musculus* has been used to demonstrate genetic variation in this enzyme (Sick and Nielson, 1964). Three phenotypes of salivary amylase could be detected; Sal–A, Sal–AB, Sal–B. The pancreatic and salivary amylases were easily separated by agar-gel electrophoresis at pH 7·3 and the pancreatic amylase also showed genetic variation. The three phenotypes of the pancreatic enzyme have been designated Pan–A, Pan–Ab and Pan–AB and are illustrated in Fig. 46. The salivary amylase appears to be controlled by two co-dominant allelic genes $Amy–1^A$ and $Amy–1^B$. The pancreatic amylases are probably controlled by a two-allele system with the A phenotype having the genotype $Amy–2^A/Amy–2^A$, $Amy–3^A/Amy–3^A$, the Ab phenotype the genotype $Amy–2^A/Amy–2^B$, $Amy–3^A/Amy–3^A$, and the AB phenotype the genotype $Amy–2^B/Amy–2^B$, $Amy–3^A/Amy–3^A$. The allele $Amy–3^A$ determines an amylase with the same electrophoretic mobility as that determined by the $Amy–2^A$ allele. No variation of the $Amy–3$ locus has been found. The $Amy–1$ and $Amy–2$ loci appear to be closely linked.

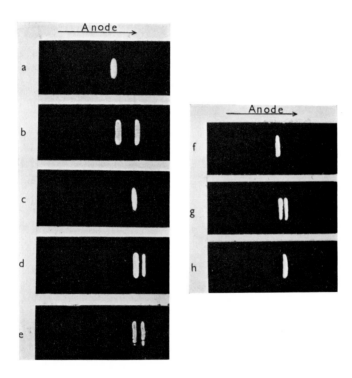

FIG. 46. Genetic variation of mouse amylases. a, Sal–A; b, Sal–A + Pan–A; c, Pan–A; d, Pan–Ab; e, Pan–AB; f, Sal–A; g, Sal–AB; h, Sal–B (reproduced with permission from Sick and Nielson, 1964).

11. LEUCINE AMINOPEPTIDASE

Starch-gel electrophoresis of extracts of human placentae has provided evidence of polymorphism of the enzyme which hydrolyses leucyl-β-

naphthylamide (Beckman *et al.*, 1966). Four zones of enzyme activity have been described and three types of pattern could be differentiated depending on variations in the relative amounts of two of these zones. These LAP types did not appear to be related to the placental phosphatase types or to some other perinatal factors, such as placental weight, birth weight or gestation time.

CHAPTER VII

Ontogenic and Phylogenic Studies

1. Ontogeny

During the development of an animal species to its adult stage, the isoenzyme pattern of a particular tissue is significantly altered. These changes occur during both foetal and neo-natal development and probably reflect the changing metabolic roles of the individual tissues. In order to attempt to shed some light on the ontogenic process, a number of studies have been made of patterns of tissue extracts during foetal and neo-natal development. The enzyme systems which have been most studied from this point of view are lactate and malate dehydrogenases and certain esterases.

A. Lactate Dehydrogenase

It has been observed that during the first weeks of extra-uterine life there is a gradual increase in activity of the faster moving lactate dehydrogenase isoenzymes of mouse, rat and guinea pig brain (Flexner *et al.*, 1960; Bonavita *et al.*, 1962). More bands have been demonstrated in extracts of foetal liver from the mouse and guinea pig than were present in extracts of the adult organ (Flexner *et al.*, 1960). It has also been shown that embryonic skeletal muscle and cardiac muscle of the rabbit contained all five bands (Vesell *et al.*, 1962a), whereas adult rabbit skeletal muscle contained only LDH-5 and adult rabbit heart only LDH-1. On the other hand, studies of foetal and neo-natal tissues in the mouse (Markert and Ursprung, 1962) have indicated that practically all tissues examined showed major bands in the LDH-4 and LDH-5 position after electrophoresis. With development, there was a gradual change towards a pattern containing more of the anodically moving isoenzymes. Mouse tissues took up to three weeks after birth to develop LDH isoenzyme patterns similar to those found in the adult tissue (Markert and Ursprung, 1962). Each organ or tissue appeared to mature at a characteristic rate.

Similar observations have been reported in relation to the rat (Fine *et al.*, 1963; Latner and Skillen, 1964; Blatt *et al.*, 1966) as shown in Fig. 47. Kanungo and Singh (1965) have studied the development of LDH isoenzymes in rat tissues during post-natal development. From 4 weeks of age to 74 weeks of age the level of brain LDH-1 increased and of LDH-5 decreased by about 30%; in the same period rat heart

LDH-1 increased to a similar extent as brain LDH-1 but heart LDH-5 decreased by about 70%.

As a result of their studies on development of LDH isoenzyme patterns in rat tissues, Fine *et al.* (1963) found changes in the kidney isoenzyme pattern which could be related to morphological development of renal tissue. Smith and Kissane (1963) in a later study used the

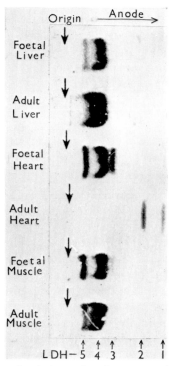

Fig. 47.   Starch-gel electrophoresis patterns of lactate dehydrogenase in foetal (2–3 days before birth) rat tissues (Latner and Skillen, 1964).

ratio of the lactate dehydrogenase activities at different pyruvate concentrations of extracts of glomeruli, aggregated cortical tubules and mesenchyme to identify the isoenzyme type predominating in these different parts of the rat kidney. Their results have confirmed that there are regional variations in LDH isoenzyme patterns. Once the glomeruli and tubules of developing kidneys are formed and recognizable they have the same isoenzyme distribution as found in the same structures in adult kidney.

Earlier studies on the lactate dehydrogenase isoenzyme patterns in human foetal tissues have shown that LDH-2, LDH-3 and LDH-4 were the predominant isoenzymes (Pfleiderer and Wachsmuth, 1961). LDH-3 and LDH-4 have been shown to be most prominent in extracts of tissues removed from a 10-week foetus (Vesell *et al.*, 1962a). Using starch block electrophoresis and relative reaction rates with various coenzyme analogues, Wiggert and Villee (1964) have confirmed that the isoenzymes with intermediate mobility were more predominant in foetal than in adult tissues. Using starch gel electrophoresis with visual demonstration of the isoenzyme patterns, Latner and Skillen (1964) and Zinkham *et al.* (1966b) have shown that in human foetal tissues the LDH isoenzyme pattern resembled that of the fully developed adult at a much earlier stage than did most other species, as shown in Fig. 48. Takasu and Hughes (1966) in a study of lactate dehydrogenase isoenzymes in

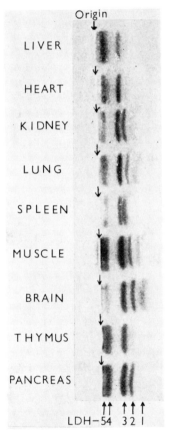

Fig. 48. Starch-gel electrophoresis patterns of lactate dehydrogenase in developing human tissues. The tissues were taken from a 14-week foetus (Latner and Skillen, 1964).

developing human muscle have stressed the importance of always comparing patterns from the same muscle, as normal adult muscles had
significantly different patterns from each other. They have shown that
the pattern of rectus femoris muscle changed markedly from the 5-month
foetus to the newborn infant and to the adult. In the foetus the isoenzymes of intermediate mobility were most prominent and during
development the pattern altered so that LDH-5 was the most predominant isoenzyme. These changes might represent changes in the
relative composition of red and white fibres or even changes in the
isoenzyme pattern of the fibres themselves.

In ontogenic studies of porcine lactate dehydrogenase, Markert and
Møller (1959) have shown differences between the patterns obtained
from adult and foetal tissues. They did, however, report on the results
from one particular foetal age only. Fieldhouse and Masters (1966)
have made a more detailed investigation of the development of LDH
isoenzymes in this species. Using starch-gel electrophoresis and relative
activities with pyruvate and α-oxobutyrate, it has been possible to show
that most porcine foetal tissues had approximately the same type of
isoenzyme pattern. During the gestation and post-natal period the pattern for skeletal muscle showed a remarkable change from one consisting
mainly of LDH-1, LDH-2 and LDH-3 during early foetal life to one
consisting mainly of LDH-5 in adult life.

Studies of the development of LDH isoenzyme patterns in ovine
tissues have also been reported (Masters, 1964) and compared with
their development in bovine tissues (Hinks and Masters, 1964). From
these investigations it has been shown that the changes in development
of the muscle lactate dehydrogenases in both species proceeded differently from those in liver and heart. In all early ovine foetal tissues
studied there appeared to be about 70% of the H or B sub-unit and in
bovine tissues about 60% of the same sub-unit. During development the
amount of H sub-unit increased to 80–85% in ovine heart and liver and
to 75–80% in bovine heart and liver, and decreased to about 30% and
15% in ovine and bovine muscle respectively.

The pattern of lactate dehydrogenase isoenzymes in the developing
chicken has also been examined (Lindsay, 1963). LDH-1 was the principal isoenzyme in the chick embryo. In the breast muscle, lactate
dehydrogenase activity gradually shifted along the spectrum of isoenzymes until the principal one present was LDH-5, as in the adult.
Similar findings using both electrophoresis and sensitivity towards coenzyme analogues have been obtained by other workers (Cahn et al.,
1962; Kaplan and Cahn, 1962; Wiggert and Villee, 1964). Nebel and
Conklin (1964) have described six LDH isoenzymes in chick embryo

tissues. Five occurred in the liver and the sixth was found in extracts of embryo brain and spleen. Only slight changes in the brain and spleen isoenzyme patterns could be detected during development. Maisel *et al.* (1965) have reported that up to nine different forms of lactate dehydrogenase were present in the chick lens. In the chick embryo one to five isoenzymes could usually be detected and between six and nine isoenzymes were found to be visible after the chicks had hatched.

Further comparative studies of the development of lactate dehydrogenase isoenzymes in tissues of the cat, pig and guinea pig (Masters and Hinks, 1966) have shown marked differences in the development of liver and heart LDH isoenzyme patterns in these species. In tissues from early foetal life, pig and cat lactate dehydrogenase was made up of about 60% of the H or B sub-unit. During development to the adult stage, the percentage of the H or B sub-unit in pig liver increased to 70% and in cat liver decreased to 30%. In the guinea pig liver the concentration of this sub-unit increased from 60% to 80% during the foetal period but then decreased to an adult level of 40%. With heart lactate dehydrogenase the percentage of the B sub-unit in the foetal cat and the foetal pig was approximately 60% and in the foetal guinea pig only 20%. During development of these animals, the level of the A sub-unit appeared to be increased to about 80%.

From all these results it would seem, therefore, that there are two possible developmental mechanisms. One in which there is a greater variety of isoenzymes in foetal life which is then cut down and another where the adult pattern develops from an initial foetal picture showing only the slower moving isoenzymes in practically all tissues. Although the results with some species indicated that early embryonic lactate dehydrogenase isoenzymes approached the pure type, i.e. LDH-1 or LDH-5, in general it appeared that the hybrid forms LDH-2, LDH-3 and LDH-4 were the most predominant in foetal tissues and that the adult tissue patterns consisting of predominantly faster or slower moving isoenzymes resulted from a gradual shift in the spectrum during foetal and neo-natal life. There were wide species variations in these developmental changes and also wide variations in the rate of development of the different tissue patterns within a particular species. Newborn rat tissues had a close resemblance to foetal tissues and the development of the adult pattern took place during the first 3 weeks or so of neo-natal growth (Fine *et al.*, 1963; Latner and Skillen, 1964). The development of the isoenzyme patterns of human heart and brain compared with rat heart and brain served to indicate the species differences in development (Latner and Skillen, 1964), as also did other comparative studies (Masters and Hinks, 1966).

The relationship between metabolic function and physiological environment and lactate dehydrogenase isoenzyme pattern has been used by some workers to explain changes during development (Fine *et al.*, 1963; Markert, 1963b). It has been proposed that the relatively low oxygen supply to mammalian foetal tissues has a bearing on the greater predominance of the slower moving isoenzymes in such tissues (Dawson *et al.*, 1965). However, the post-natal changes in rat tissue isoenzyme patterns which are well documented (Fine *et al.*, 1963; Latner and Skillen, 1964; Kanungo and Singh, 1965; Blatt *et al.*, 1966) cannot really be explained satisfactorily on this basis unless it can be shown that there is a marked change in the rate of cellular differentiation during the first two to three post-natal weeks. A relationship between mitosis and the distribution of human heart lactate dehydrogenase isoenzymes has been described (Wachsmuth, 1964). During the mitotic phase, lactate dehydrogenase isoenzymes of intermediate mobility predominated, whereas in the post-mitotic state, tissues had attained their normal isoenzyme patterns. These inferences have been made from studies of the ventricles of human heart; the left ventricle has been shown to develop the characteristic isoenzyme pattern of adult human heart much earlier than the right ventricle and this has been correlated with the earlier onset of the amitotic phase in the former.

B. LACTATE DEHYDROGENASE DURING ONTOGENESIS IN FISH, MARINE ANIMALS AND AMPHIBIA

An ontogenic study of lactate dehydrogenase isoenzymes in the prosobranch mollusc *Argobuccinum oregonense* has produced some interesting findings (Goldberg and Cather, 1963). In contrast to mammalian embryonic tissues, it was found that development in this mollusc started with two isoenzymes of exceptionally high electrophoretic mobility in addition to some other isoenzymes with the more usual mobility. During development the two fast moving isoenzymes disappeared with a corresponding increase in the normal LDH-1 and LDH-3.*

The development of LDH isoenzyme patterns has also been studied in two species of amphibians (Adams and Finnegan, 1965). No increase in the number of isoenzymes after starch-gel electrophoresis could be detected during the early developmental stages of the salamander, *Amblystona gracile*, ten to twelve bands being present at all stages. In the post-neurulation stages of development, however, there appeared to be some losses in both the fastest and slowest moving entities. Using starch-gel electrophoresis, only a single zone of lactate dehydrogenase could

---

* A reduction in the number of malate dehydrogenase components during development of *Arbacia* has been demonstrated (Moore and Villee, 1963).

be detected in *Rana aurora*, except at the stage when autonomous movement was evident when a second isoenzyme was found (Adams and Finnegan, 1965). Disc electrophoresis in acrylamide gels yielded different results from those with starch gels (Adams and Finnegan, 1965). The number of isoenzymes demonstrated in *A. gracile* was much reduced, especially in the early stages of development, but in *R. aurora* three to four components were detected at nearly all stages.

Studies of the changes in LDH during thyroxine-induced tadpole metamorphosis have shown a single enzyme component in tadpole liver, with three in tadpole brain and tail (Kim *et al.*, 1966). Treatment with thyroxine produced a decrease in the LDH activity of all three organs but this decrease was confined to only one of the three isoenzymes present in brain and tail.

The changing isoenzyme patterns of a number of dehydrogenases in developing tissues of the cyprinodontiform fish *Oryzias latipes*, the medaka, have been examined using disc electrophoresis (Nakano and Whiteley, 1965). Only a single form of lactate dehydrogenase (LDH-5) could be detected before hatching but four others appeared abruptly on hatching. All adult organs have been shown to possess one to three isoenzymes. LDH-1 and LDH-2 were found only in the retina.

C. CREATINE KINASE

The development of creatine kinase isoenzyme patterns in extracts of rat brain, cardiac and skeletal muscles, has been studied with agar-gel electrophoresis (Eppenburger *et al.*, 1964). In adult skeletal muscle a single slow moving component has been found, whereas in foetal rat skeletal muscle, 8–10 days before birth, a single fast moving component was present. During foetal and neo-natal development, a gradual change occurred in the isoenzyme spectrum from the fast moving entity to one of intermediate mobility and finally to the slow moving isoenzyme. At birth all three were present and the full adult pattern was not reached earlier than 30 days after birth. These changes are shown in Fig. 49. In cardiac muscle, the foetal pattern appeared as the fast moving isoenzyme with traces of the one with intermediate mobility. At birth all three isoenzymes could be detected and the mature adult pattern showed approximately equal amounts of the isoenzymes of slow and intermediate mobility. It appears that in rat cardiac muscle, development of the adult pattern begins earlier than in skeletal muscle but adult pattern development does not require such marked changes as skeletal muscle. Rat brain had essentially the same pattern in foetal and adult tissues.

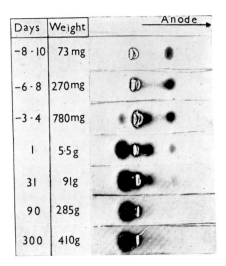

F<small>IG</small>. 49.   Agar-gel electrophoresis patterns of creatine kinase in developing rat muscle, the rats being born on day 0 (reproduced with permission from Eppenburger *et al.*, 1964).

2. E<small>VOLUTIONARY AND PHYLOGENIC</small> S<small>TUDIES</small>

Temperature stability of the H-type lactate dehydrogenases has been shown to vary markedly in different species (Wilson *et al.*, 1964). Whereas the activity of the H_4 isoenzyme from man was reduced by 50% by heating to 65°C for 20 minutes, the activities of the enzymes from the lizard *Iguana iguana*, the ostrich, the bullfrog and the mackerel were reduced by 50% by heating for 20 minutes at 82°C, 80°C, 52°C and 60°C respectively. The results with temperature stability techniques have indicated that at one point in vertebrate evolution there was a very marked change. All lower vertebrates, such as fish and amphibia, had temperature stability half-lives for H_4 around about 60°C, whereas the higher reptiles and many birds had values around about 80°C.

Studies of the lactate dehydrogenase isoenzyme patterns in brain extracts of eleven vertebrate species have shown that there was a consistent phylogenic sequence through the series (Bonavita and Guarneri, 1963a). The electrophoretic mobility of the most prominent isoenzyme in the brain extract increased as one went up the phylogenic series from a Selachian *Mustelus mustelus* to the cat, *Felis catus* (Fig. 5). Corresponding to these changes in electrophoretic mobility were variations in the kinetic characteristics of the enzymes from different phyla. Enzymes from brains of lower phyla had higher K_m values for lactate than those from brains of higher phyla. Pyruvate inhibition has been found to be much more significant for enzymes from mammals than from fish. Relative reaction rates with coenzyme analogues and bisulphite inhibition have also provided evidence of the phylogenic alterations in isoenzyme patterns.

In twenty-one species of neognathous birds the electrophoretic mobility of the H_4 isoenzyme was approximately 2 cm in 16 hours at 10 V/cm; in ten species of reptiles the mobility was approximately 6 cm; in eleven species of mammals it was about 15 cm and in twelve widely differing species of amphibia and fish it was less than 10 cm (Wilson *et al.*, 1964); in three species of paleognathous birds the value was round about 7 cm. These findings, in combination with heat stability properties, fit very closely with what is known of evolutionary pathways, as shown in Fig. 50. In fishes the H_4 isoenzyme is unstable

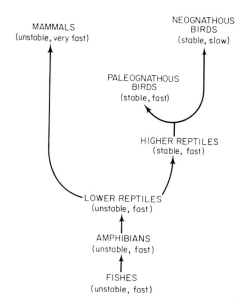

Fig. 50.   Evolutionary tree showing the changes in heat stability and mobility of lactate dehydrogenase (reproduced with permission from Wilson *et al.*, 1964).

to heat and relatively fast moving; in amphibia and lower reptiles the isoenzyme is unstable to heat and fast moving; in higher reptiles the enzyme is stable to heat and fast moving; in paleognathous and neognathous birds the enzyme is respectively heat stable and fast moving and heat stable and slow moving. In mammals the enzyme is heat unstable and very fast moving (Wilson *et al.*, 1964).

F*

CHAPTER VIII

Insects, Plants and Lower Organisms

1. INSECTS

ISOENZYME studies in insects have been largely concerned with genetic variations. The majority of these studies have been applied to *Drosophila* and will be discussed separately.

A. STUDIES WITH DROSOPHILA

(i) *Alcohol dehydrogenase*. Isoenzymes of alcohol dehydrogenase in extracts of *Drosophila melanogaster* were first described by Johnson and Denniston (1964). Two isoenzymes were found in each true breeding stock and five in hybrid flies. Confirmation of the existence of isoenzymes has been obtained by Grell *et al.* (1965) and Ursprung and Leone (1965). The former group of workers used acrylamide-gel electrophoresis and found that the enzyme existed in three electrophoretically different forms in inbred strains of *D. melanogaster*. There were two types of pattern, a slow type which has been detected in Canton-S wild-type flies and a fast type which has been detected in Samarkand, Swedish-6 and many other wild-type flies. Nine zones of enzyme activity have been visualized after acrylamide-gel electrophoresis of hybrids between the two types. Of these nine zones, 1, 4 and 7 belonged to the slow-type parent, zones 3, 6 and 9 to the fast-type parent, whereas 2, 5 and 8 were specific for the hybrid (Grell *et al.*, 1965). It has been suggested that the enzyme exists as a dimer of two polypeptide sub-units (Grell *et al.*, 1965). Further genetic studies have also indicated that the gene for alcohol dehydrogenase is located on the second chromosome with a map position of 50·1 and a cytological position between 34E3 and 35D1. Using agar-gel electrophoresis, it has been possible to confirm the existence of three zones in homozygous stocks of flies (Ursprung and Leone, 1965). Only seven zones of enzyme activity could be detected after agar-gel electrophoresis of extracts of hybrid flies. This has been attributed to the better resolution provided by acrylamide-gel electrophoresis. Homozygous flies of Types I and II contained three isoenzymes, two of which were common to both types. Heterozygotes, Type III, contained seven isoenzymes, i.e. four parental isoenzymes and three hybrid isoenzymes (Fig. 51). The proposed explanation of these findings is that Type I flies synthesize two sub-units 1^A and 2^A which aggregate in the form of dimers to give 1^A1^A, 1^A2^A and 2^A2^A. Type II flies apparently synthesize two sub-units 1^B and 2^B and aggregate in a similar manner to give 1^B1^B, 1^B2^B and 2^B2^B. Type III flies, the heterozygotes, should contain ten isoenzyme dimers, viz. 1^A1^A, 1^A1^B, 1^B1^B, 2^A2^A, 2^A2^B, 2^B2^B, 1^A2^A, 1^B2^B, 1^A2^B and 1^B2^A. As three pairs 1^A2^A and 1^B1^B, 1^A2^B and 1^B2^A and 2^A2^A and 1^B2^B have similar mobilities only seven isoenzymes can be detected after agar-gel electrophoresis (Ursprung and Leone, 1965). An octanol dehydrogenase with polymorphism has also been demonstrated in different stocks of *D. melanogaster* (Ursprung and Leone, 1965).

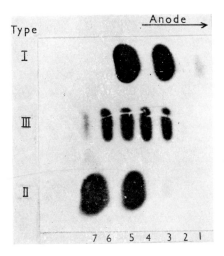

Fig. 51.   Alcohol dehydrogenase in *Drosophila* (reproduced with permission from Ursprung and Leone, 1965).

(ii) *Xanthine dehydrogenase.* Following preliminary studies with paper electrophoresis (Keller *et al.*, 1963) it has been possible to demonstrate three zones of enzyme activity after gel electrophoresis of extracts of *Drosophila* (Glassman and Saverance, 1963; Smith *et al.*, 1963). Up to three zones have been visualized in extracts of a wild-type strain of *D. melanogaster* although only one of these zones could be definitely identified as xanthine dehydrogenase. Although no differences in the mobility of the major known xanthine dehydrogenase component could be detected in the various strains of *D. melanogaster* examined, differences have been found between *D. melanogaster* and *D. virilis* (Glassman and Saverance, 1963). It has been suggested (K. D. Smith *et al.*, 1963) that the isoenzymes may be composed of three or four polypeptide chains, the synthesis of which is controlled by two genes. Further investigations have enabled Yen and Glassman (1965) to demonstrate four types of electrophoretic variants of xanthine dehydrogenase in fifty-six strains of *D. melanogaster*. These four variants have been termed slow (S), slow intermediate (SI), intermediate (I) and fast (F). Crosses between flies of the different types produced hybrid molecules with mobilities intermediate between those of the parents (Yen and Glassman, 1965).

(iii) *L-leucyl-β-naphthylamidase*. Six zones of arylamidase activity have been detected following starch gel electrophoresis of homogenates of pupae of *D. melanogaster* (Beckman and Johnson, 1964a). They have been termed A, B, C, D, E and F. All except zone A were apparently specific for the pupal stage of life. There were differences in the patterns from early and late pupae, as the slower moving zones were present only towards the end of pupal life.

Genetic variants of both the Lap A and Lap D zones have been described. That of the D zone of *D. melanogaster* appeared to be controlled by a pair of co-dominant autosomal alleles, one Lap D^F controlling synthesis of the fast moving variant, and the other, Lap D^S, the slow moving variant. The homozygotes possessed either a fast or slow moving variant of the enzyme, whereas in heterozygote flies both zones could be detected in equal amounts; there was no indication of a hybrid enzyme (Beckman and Johnson, 1964a). The variants of the A zone have been shown to be related to the inheritance of the Lap D enzyme. Pupae which were homozygous for the Lap D^F allele contained a slow moving A isoenzyme; those homozygous for the Lap D^S allele, a fast moving A isoenzyme whereas the heterozygotes Lap D^F/Lap D^S, contained the Lap A isoenzyme of intermediate mobility. The Lap A zones have been found to be absent in quite a large number of flies which were homozygous for the Lap D^F or Lap D^S alleles. This absence has been shown to be inherited as a recessive trait. No evidence has been found for recombination between Lap A and Lap D genes. It has been suggested that the Lap A and Lap D isoenzymes are controlled by two independent closely linked loci.

Polymorphism has also been described in *D. buskii* (Johnson and Sakai, 1964), although the patterns differed markedly from those previously found for *D. melanogaster*. Those of *D. buskii* appeared to be controlled by three co-dominant alleles. Flies homozygous for the so-called Lap^F, Lap^M and Lap^S alleles contained a single isoenzyme of fast, intermediate or slow mobility respectively. Heterozygotes contained a pair of isoenzymes each having the same mobility as those of the respective parents.

(iv) *Phenol oxidases.* The phenol oxidase systems of *Drosophila melanogaster* have been studied by means of acrylamide gel electrophoresis (Mitchell and Weber, 1965; Geiger and Mitchell, 1966; Mitchell, 1966). At least four protein components have been shown to be involved; one (A1) plus another (P) yielded tyrosinase while components A1 and A2 together yielded 3,4-dihydroxyphenylalanine oxidases (Mitchell and Weber, 1965). It has been suggested that the four components may possibly represent specialized sub-units (Mitchell and Weber, 1965). Further studies on the development of these systems in *Drosophila* have been described (Mitchell, 1966; Geiger and Mitchell, 1966).

(v) *Amylase.* Following agar-gel electrophoresis of extracts of *D. melanogaster*, seven zones of amylase have been detected, whereas with *D. virilis* only two zones were present (Kikkawa, 1963). The various strain differences in the amylase isoenzymes and the polymorphism within particular inbred strains appeared to be controlled by allelic genes on the second chromosome (Kikkawa and Ogita, 1962).

(vi) *Alkaline phosphatase.* Starch gel electrophoresis has been used in studies of the alkaline phosphatases of *Drosophila melanogaster* (Beckman and Johnson, 1964b). Three variants of the enzyme have been described, one a fast moving zone, the second a slow moving zone and the third a zone of intermediate mobility. The inheritance of the enzyme appeared to be controlled by a pair of codominant allelic genes located on the third chromosome. Heterozygote flies showed the parental isoenzymes in addition to a hybrid band.

As well as showing genetic variations, the phosphatase patterns varied during the developmental stages of the *Drosophila* larvae (Beckman and Johnson, 1964b). The varying patterns of alkaline phosphatase isoenzymes in developing *Drosophila* have been confirmed by Schneiderman and co-workers (1966) who have demonstrated a total of seven different components. Their appearance and non-appearance during development, from young eggs to larvae, to pupae and finally to mature flies, has been related to organ-specificity.

A deficiency of a normally prominent alkaline phosphatase in larvae of *Drosophila melanogaster* has been detected in some cases (Johnson, 1966). This deficiency appeared to be under the control of an allele concerned in the previously established polymorphism. It has been suggested that this newly described allele is related to the synthesis of an enzymatically inactive sub-unit which is able to hybridize with the sub-units of the slow electrophoretic variant (Johnson, 1966).

A comparison of the larvae, pupae and adult flies of *D. melanogaster* with other species of this insect has shown that the phosphatases of *D. simulans* were identical in pattern and mobility, whereas those of species belonging to another sub-genera such as *D. virilis* and *D. funebris* were somewhat different (Schneiderman *et al.*, 1966).

Differences in the alkaline phosphatase zymograms during development of *D. melanogaster* and *D. ananassae* may be due to addition or substitution of a more highly charged enzyme sub-unit (Johnson, 1966). In some males of *D. ananassae* a reversal to the larval type of alkaline phosphatase pattern might have represented possible reactivation of synthesis of an enzyme sub-unit.

(vii) *Acid phosphatases*. Acid phosphatase has been described as another gene-enzyme system in the sibling species *D. melanogaster* and *D. simulans* (MacIntyre, 1966). In both species the genes appeared to be located on the third chromosome near to the homologous claret genes. The homologous nature of the former genes was supported by the existence of typical heterozygous patterns in species hybrids. The isoenzymes in both species appeared to be controlled by two codominant interacting alleles at a single autosomal locus (MacIntyre, 1966).

(viii) *Phosphogluconate dehydrogenase*. Certain strains of *Drosophila* contain phosphogluconate dehydrogenases showing varying mobility during starch-gel electrophoresis (Kazazian *et al.*, 1965). There is usually either a single zone designated PGD A or PGD B, or three zones PGD A, PGD B, PGD I[AB] where the latter is the most intense and of intermediate mobility. The isoenzyme of this zone could be produced *in vitro* by dissociation and recombination of the isoenzymes in zones PGD A and PGD B using propanedithiol. The gene *Pgd* appeared to be located on the X-chromosome (Kazazian *et al.*, 1965).

(ix) *Glucose-6-phosphate dehydrogenase*. Three types of pattern of glucose-6-phosphate dehydrogenase have been described after starch-gel electrophoresis of extracts of wild-type strains of *D. melanogaster* (Young *et al.*, 1964). Of the nineteen strains examined, five showed a slow moving isoenzyme, twelve a fast moving isoenzyme and two showed both. Genetic studies have indicated that the enzyme polymorphism was controlled by a gene located on the X-chromosome (Young *et al.*, 1964).

(x) *Esterases*. Following starch-gel electrophoresis of extracts of *D. melanogaster*, ten zones of non-specific esterase activity have been detected (Wright, 1963). The most clearly identifiable components were the so-called Est 6 esterases which occurred as either a fast (Est 6[F]) or slow (Est 6[S]) moving component in homozygous flies. Both components were present in heterozygous flies. The two isoenzymes have been shown to differ in their sensitivities to heat and organophosphorus inhibitors, as well as in electrophoretic mobility (Wright, 1963). These Est 6 enzymes appear to be under the control of a pair of codominant alleles located on the third chromosome (Wright, 1963). Further studies have indicated that in *D. melanogaster* and *D. simulans* the two pairs of alleles controlling the Est 6 enzymes constitute an homologous gene–enzyme system (Wright and McIntyre, 1963).

Beckman and Johnson (1964c) have also studied esterase variations in *D. melanogaster* and have found six different esterase components (A–F). The fastest moving zone during starch-gel electrophoresis (A) was found to be best developed in pupae and showed some variations in mobility not necessarily genetically determined. The second fastest moving zone (B) was rather faint and was not detected in all individuals. Zones C or D were present in larvae, pupae and adult flies and showed electrophoretic variations, the D zone being identical with the Est 6 esterases of Wright (1963). Band E was probably identical with cholinesterase as it was inhibited by 10^{-5} M eserine. Zone F has been tentatively identified as an ali-esterase (Beckman and Johnson, 1964c). The C zone consisted of either one or two components. Either the fast or slow moving isoenzyme has been detected in homozygous flies. Heterozygotes contained both (Beckman and Johnson, 1964c). The Est C enzymes appeared to be under the control of two autosomal codominant alleles. Genetic studies using the Est 6 alleles as markers have indicated that the locus for each of these alleles was also on the third chromosome.

The responses of the Est 6 alleles to selection in an experimental population of *D. melanogaster* and *D. simulans* have been examined and it has been found that where no attempt was made to control the genetic background, equilibrium frequencies were established in less than ten generations (MacIntyre and Wright, 1966). This situation has been found to hold even when the founders were from the same or different stocks and both alleles appeared to persist at rather stable relative proportions in most populations.

B. STUDIES WITH OTHER INSECTS

In the silkmoths Cynthia (*Samia cynthia*) and Cecropia (*Hyalophora cecropia*) it has been shown by means of starch-gel electrophoresis that a number of hydrolases and oxidoreductases exist in multiple forms in their blood and tissues (Laufer, 1960). At least eight forms of nonspecific esterases have been detected in *Cecropia* blood, although the number of esterases varied with the stage of development from pupal to adult life (Laufer, 1960; 1961). They appeared to have distinct substrate specificities and indications have been obtained of tissue specific esterase patterns (Laufer, 1960; 1961). Studies on the lactate, malate and α-glycerophosphate dehydrogenases have also provided evidence that changes in tissue specific isoenzyme patterns during development might be under hormonal control (Laufer, 1961).

Immunochemical and electrophoretic investigations of the esterase isoenzyme patterns of the blowfly *Phormia regina* have also shown tissue specificity and alteration during development (Laufer, 1963).

Studies on the housefly (*Musca domestica*) have shown the presence of allelic genes modifying esterase enzymes so that they were resistant to organo-phosphates (Oppenoorth and Van Asperen, 1960). Using agar-gel electrophoresis seven esterase components have been demonstrated in extracts of *Musca domestica* (Van Asperen, 1962; 1964). Evidence of strain differences in the esterase isoenzyme patterns has been described. Further studies of the electrophoretic properties of the esterases from susceptible and resistant strains of *Musca domestica* have shown that they can be split into the three groups, cholinesterases, aliesterases and aromatic esterases, according to their sensitivity to the inhibitors 10^{-4} M eserine, 10^{-4} M paraxon and 10^{-4} M DDVP. Four cholinesterase, two aliesterase and three aromatic esterase components have been described (Menzel *et al.*, 1963).

In certain strains of the housefly, ten different esterase components have been described, individual flies containing three to eight of these. Genetic studies of eleven strains of housefly have shown that the esterases are under the control of nine genes on eight different chromosomal loci (Velthuis and Van Asperen, 1963; Van Asperen and Van Mazijk, 1965). It has been demonstrated that resistance to organo-phosphate insecticides was related to the absence of an aliesterase component which, in the susceptible strain, represented 85% of the total non-eserine sensitive esterase activity.

Multiple forms of esterases are apparently present throughout the insect kingdom. Using column chromatography on DEAE cellulose, Arfsharpour and O'Brien (1963) have studied the enzymes which hydrolyse α-naphthyl acetate, indophenol acetate and 5-bromoindoxyl acetate in the American cockroach, Colorado potato beetle, spider mite, honeybee, housefly, American bean beetle, and the milkweed bug. Changes in the esterase isoenzyme patterns of some of these species have been shown to occur during development.

Gilbert and Huddlestone (1965) have been able to demonstrate multiple forms of acid phosphatase in the testes of the giant silk-moth. Two components could be visualized in *Samia cynthia* and three in *Hyalophora cecropia*.

The evidence for multiple forms of inorganic pyrophosphatase in the boll weevil, *Anthomus grandis*, was not conclusive. Of the six zones detected after paper electrophoresis, only one (85%) had more than 10% of the total enzyme activity (Lambremont and Schrader, 1964).

2. Higher plants

A. isoenzymes in maize

(i) *Esterases*. Using starch gel electrophoresis, genetic studies of hybrid enzymes in maize endosperm extracts have shown three forms of esterase which are under the control of three allelic genes (Schwartz, 1960). Homozygotes showed either a fast or slow moving isoenzyme and the heterozygotes showed a hybrid isoenzyme of intermediate mobility, as well as the two parental types. Further studies (Schwartz, 1964a; Schwartz *et al.*, 1965) have shown that there were seven alleles of the so-called E_1 esterase gene. The original genes $E_1{}^F$, $E_1{}^N$ and $E_1{}^S$ occurred in North American maize, and other genes $E_1{}^L$, $E_1{}^R$ in teosinte and $E_1{}^T$ and $E_1{}^W$ in a strain of maize from South America. These seven genes apparently control seven autodimer enzymes, each with a different electrophoretic mobility. All twenty-one possible heterozygotes have been produced and in each case a hybrid enzyme has formed in addition to the parental types (Schwartz *et al.*, 1965). It has been demonstrated that the migration rates of the isoenzymes in the two series E^F, E^N, E^S and E^L, E^R, E^T and E^W overlapped and that identical charge differences distinguished the isoenzymes in both series (Schwartz *et al.*, 1965). This is shown in Fig. 52. Treatment of the maize esterases with borohydride

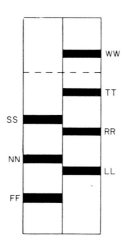

Fig. 52.   Maize esterases (reproduced with permission from Schwartz *et al.*, 1965).

has shown that a single isoenzyme can be converted into a number of new enzyme forms, all more negatively charged than the original isoenzyme (Schwartz, 1964b). It has been suggested that this interconversion of allelic isoenzymes is due to variation in the number of charged conjugated groups rather than differences in the net charge of amino-acid components. Incubation of the so-called pH 7·5 maize esterase with glyceraldehyde resulted in the conversion of the enzyme to an acidic form. As this was a gradual process the enzyme could be stabilized in intermediate forms (Schwartz, 1965). This treatment eliminated all the genetically determined charge differences that distinguished the seven isoenzymes and all were converted to the same acidic form.

(ii) *Alcohol dehydrogenase*. Polymorphism of maize alcohol dehydrogenase has also been described (Schwartz and Endo, 1966). Three allelic forms have been detected after starch-gel electrophoresis of the scutellum of the mature kernel. The alleles have been designated Adh^S, Adh^F and Adh^C where S indicated the slowest and C the fastest migrating isoenzyme. Each isoenzyme was apparently a dimer and in plants homozygous for each of the alleles a single zone was obtained. In heterozygote plants, a hybrid enzyme with intermediate migration rate was observed in addition to the two parental enzymes. The enzyme controlled by a cross between the Adh^C and Adh^S alleles had the same mobility as the homozygote Adh^F/Adh^F. A further compound locus has been described which contains two alleles Adh^F and Adh^C. Crosses between Adh^{FC} and Adh^F showed three isoenzymes FF, FC and CC and those between Adh^{FC} and Adh^S showed five isoenzymes SS, SF, FF/CS, CF and CC. Investigations have indicated that the Adh^C alleles might exist in two forms $Adh^{C(m)}$ and $Adh^{C(t)}$ where m and t represent maize and teosinte respectively. The isoenzymes controlled by these alleles have identical migration rates, but different specific activities.

(iii) *Arylamidase*. The leucyl β-naphthylamidase of maize has also been investigated (Beckman *et al.*, 1964a; Scandalios, 1964; 1965). Four zones of the enzyme have been detected after electrophoresis of maize kernel extracts. These have been termed A, B, C and D in order of decreasing mobility towards the anode. The B and C enzymes have been separated very rarely in extracts from leaves and husk. In seedlings the D enzyme was absent. Tissue specificity of the naphthylamidase, as well as esterases, peroxidases and catalases has also been described by Scandalios (1964) who has reported on the changes in the naphthylamidase isoenzyme pattern during development of maize endosperm, pollen, young ear, silks, husk, root, stem and leaf (Scandalios, 1965).

Genetic variation in both the A and D enzymes has been described. Each isoenzyme is most probably a dimer, since homozygotes for both alleles showed either a single fast or slow moving zone, whereas heterozygotes possessed a hybrid enzyme, as well as the parental enzymes. Back-cross data have indicated that the genes controlling the variations in both the A and D enzymes were closely linked (Beckman *et al.*, 1964a).

(iv) *Catalase*. Five catalase isoenzymes have been detected in extracts of maize endosperm (Beckman *et al.*, 1964b). The results have indicated that the enzyme is analogous to mammalian lactate dehydrogenase in so far as it apparently exists as a tetramer with three hybrid enzymes. The three types are under genetic control, the homozygotes possessing only a fast or slow moving isoenzyme and the heterozygotes five isoenzymes. By freezing and thawing a mixture of the slowest and fastest moving isoenzymes in a solution of molar sodium chloride, the five isoenzymes were produced.

B. ISOENZYMES IN OTHER PLANTS

(i) *Oxidoreductases*. The multiple forms of glutamate dehydrogenase in plants such as the broad bean *Vicia fabia*, and the pea *Pisum sativum* have been examined (Thurman *et al.*, 1965). A total of seven zones have been detected and there were changes in the number and distribution during germination; in other words the glutamate dehydrogenase isoenzyme pattern was relatively tissue specific for the cotedylons, radicles and shoots of these plants. Three zones of glutamate dehydrogenase and a single zone of isocitrate dehydrogenase have been reported after electrophoresis of root nodules of *Trifolium repens*, *Pisum sativum* and *Medicago sativa* (Grimes and Fottrell, 1966).

A preliminary study of the dehydrogenase isoenzyme patterns of legume root nodules has shown the presence of multiple forms of lactate, malate and glucose-6-phosphate dehydrogenases (Fottrell, 1966). The nodules from different species displayed characteristic dehydrogenase isoenzyme patterns but their physiological significance has yet to be determined.

Six peroxidase components have been demonstrated after starch-gel electrophoresis of corn-leaf sheath preparations (McCune, 1961). Four cathodic fractions had different kinetic characteristics. The relative proportions of each of these six fractions altered during maturation. Differences in pattern could be detected between normal plants and dwarf mutants. Giberillic acid could also be used to produce pattern changes identical with those during growth and also altered the dwarf mutant pattern to the normal one.

Up to six components have been visualized after starch-gel electrophoresis of commercial preparations of horse-radish peroxidase. The heterogeneity depended on charge, rather than molecular size (Klapper and Hackett, 1965).

Studies on the mutual interaction between peroxidases and the plant growth hormone, indole-3-acetic acid, have shown that young stem sections of dwarf peas grown in light yielded at least seven peroxidase isoenzymes after starch-gel electrophoresis (Ockerse *et al.*, 1966). As the tissue elongated in excised segments or in the intact plant, an eighth peroxidase isoenzyme has been detected. The appearance of this isoenzyme in excised segments could be repressed by application of the growth hormone.

Characteristic peroxidase isoenzyme patterns have been described in the root nodules of various leguminous plants, the most outstanding species differences being between *Vicia sativa* and *Galega officinalis* (Moustafa, 1963).

(ii) *Hydrolases*. A preliminary characterization of the carboxylic ester hydrolases of a number of plants has been attempted by Van der W. Jooste and Morland (1963). These workers have shown that there are complex systems of ester hydrolysis in a wide range of plants such as wheat seeds, cucumber, soybeans and corn seedlings. Comparative studies of the esterases of various species of *cucurbitaceae* have revealed the changes in esterase isoenzyme patterns during development of the stems, leaves and fruits (Schwartz *et al.*, 1964).

Agar-gel electrophoresis of the starch degrading enzymes in germinating barley seeds has shown the presence of two β-amylases, five α-amylases and two enzymes which have not yet been identified (Frydenberg and Nielson, 1965). Heat treatment of the seed extract before electrophoresis has resulted in conversion into one or two stable isoenzyme forms depending on the variety of barley. Ninety-eight barley varieties have been studied and a pedigree of inheritance of the one or two amylases presented (Frydenberg and Nielson, 1965).

Partial purification has been achieved for two DNAase/RNAase fractions and two separate RNAase fractions from extracts of germinating garlic (Carlsson and Frick, 1964).

Two acid phosphatases have been detected after acrylamide-gel electrophoresis at pH 5·2 of leaf extract of the pinto bean, *Phaseolus vulgaris* (Staples *et al.*, 1965). Both phosphatases had the same mobility during electrophoresis at pH 7·5. The two phosphatases had different heat stabilities in acetate buffer at pH 5·2 but not in tris buffer at pH 7·5.

The species differences and developmental changes in the root nodule esterase and phosphatase isoenzyme patterns have been shown to be characteristic for four different species of *Lotus* (Moustafa, 1964).

During studies on the enzymic browning of apples, Walker and Hulme (1965) have been able to extract two phenolases from apple peel. These phenolases have similar substrate specificities but differ in their behaviour on ion-exchange chromatography. Continuous flow electrophoresis of extracts of potato tubers has shown the presence of multiple forms of phenolase with two major components (Patil *et al.*, 1963); both had cresolase and catecholase activity. The heterogeneity of tyrosinase in the broad bean *Vicia faba* has been demonstrated (Robb *et al.*, 1965).

C. CHANGES IN DISEASE

Ceratocystic infection in sweet potatoes has been shown to alter the peroxidase isoenzyme patterns of the plant tissues (Weber and Stahmann, 1964).

Changes in the protein and isoenzyme patterns of susceptible bean leaves after infection by the bean rust fungus have been described (Staples and Stahmann, 1964). Increased numbers of isoenzymes of acid and alkaline phosphatase and succinate and malate dehydrogenases have been detected after acrylamide-gel electrophoresis of extracts of *Phaseolus vulgaris* L. var. *Pinto* infected with *Uromyces phaseoli*. Changes in some protein and enzyme patterns of sweet potato roots in response to black rot infection have been examined, using chromatography on DEAE cellulose (Kawashima *et al.*, 1964). Disc electrophoresis of tissues of *Phaseolus vulgaris* infected with *Pseudomonas phaseolica* (Halo blight of bean) showed the presence of additional catalase and peroxidase components in the diseased tissue when compared with normal tissue (Rudolph and Stahmann, 1964).

3. FUNGI

Multiple forms of phenolases have been separated from extracts of mushrooms by means of absorption on calcium phosphate gel and chromatography on hydroxyapatite and DEAE-cellulose. Heterogeneous peaks corresponding to catecholases and cresolases have been obtained (Smith and Krueger, 1962). Four components have been separated by continuous flow electrophoresis. They had differing activities towards mono- and ortho-diphenols (Bouchilloux *et al.*, 1963). Using starch-gel and acrylamide-gel electrophoresis, Jolley and Mason

(1965) have been able to show that the peaks obtained by Bouchilloux and co-workers (1963) were heterogeneous; up to five or six zones being detectable by a visual staining technique. It would appear that these multiple phenolases from mushrooms are to some degree interconvertible depending upon pH, ionic strength and protein concentration and the existence of isoenzymes may be due to varying degrees of polymerization or different combinations of unlike sub-units (Jolley and Mason, 1965).

Two forms of tyrosinase have been demonstrated in a crystalline preparation of the enzyme from *Neurospora crassa* (Fling *et al.*, 1963). The component with the slightly greater electrophoretic mobility was also the more heat labile.

Crystalline yeast hexokinase has been separated into six components by chromatography on DEAE cellulose (Kaji *et al.*, 1961). Two of these fractions had identical kinetic characteristics and could be interconverted by the action of trypsin or chymotrypsin in the presence of glucose. Two forms of fumarase have been separated from extracts of the yeast *Candida utilis* and it has been shown that the two isoenzymes have different kinetic characteristics (Hayman and Alberty, 1961).

A number of alkaline phosphatase mutants of *Aspergillus nidulans* have been studied by starch-gel electrophoresis (Dorn, 1965). Three zones could be detected in wild-type strains. The distribution of total activity among the three isoenzymes was different in the temperature-sensitive mutants. The results have suggested that the phosphatases may be made up of two or more different polypeptide chains, some of which are common to two or more of the isoenzymes (Dorn, 1965).

Fincham (1962) has described multiple forms of NADP-linked glutamate dehydrogenase in *Neurospora crassa*. Mutant strains have been shown to have enzymes with different kinetic properties but which gave identical patterns after starch-gel electrophoresis. Further studies have allowed the enzymes from two of the mutants to be partially separated by chromatography on DEAE-cellulose (Fincham and Coddington, 1963). These purified enzymes can interact by mixing, adjusting to pH 5·8 and readjusting to pH 7·4. The product of this interaction is similar to that of the heterocaryon, and it is probable that this complementation represents exchange of sub-units between the interacting proteins. A mutant enzyme has been isolated from certain strains of *Neurospora crassa* which exists in two forms, one active and the other inactive (Sundaram and Fincham, 1964). These two forms are interconvertible, the active form being changed to the inactive form by briefly heating to 60°C. The inactive enzyme could be reactivated by incubation with EDTA or succinate and also, but to a lesser extent, by phosphate or citrate. Hybridization of these mutant enzyme forms is obtained by freezing and thawing in 0·05 M NaCl or by adjusting to slightly acid pH as before (Coddington *et al.*, 1966).

During studies of the proteolytic enzymes of *Aspergillus oryzae*, Bergkvist (1963) has separated three proteases on columns of DEAE cellulose after previous adsorption-elution from CM-cellulose. The three proteases had different pH optima and showed varied sensitivity towards inhibitors.

The study of multiple esterase and phosphatase components after starch-gel electrophoresis of culture filtrates of *Fusarium oxysporum* and *Fusarium xylanoides* has indicated taxa specific patterns (Meyer *et al.*, 1964).

4. PROTOZOA

The existence of multiple esterases in the protozoan *Tetrahymena pyriformis* has been established by means of starch-gel electrophoresis (S. L. Allen, 1960; 1961). They have been split into two classes on the basis of substrate specificity and sensitivity to eserine sulphate and sodium taurocholate. Up to twenty zones of non-specific esterase have been detected and those inhibited by 10^{-4} M eserine and activated by 10^{-2} M taurocholate have been observed to form two groups, one (B) with four or five components migrating towards the cathode and the other (C) with four components towards the anode at pH 7·5 (S. L. Allen, 1961). The other 10^{-4} M eserine insensitive esterases had widely varying substrate specificity. The B group of isoenzymes appeared in three inbred strains, whereas the C group was apparently confined to a single strain. Crosses between strains possessing the B and C groups have been shown to produce hybrids with both groups. It has been suggested that the two groups of isoenzymes are controlled by alleles at a single locus (S. L. Allen, 1961); the homozygote possesses only the B or C group of isoenzymes and the heterozygote has the potential of expressing both groups until it becomes differentiated so that it expresses either the B or C group. Further observations on the changes in esterase isoenzyme patterns during the growth cycle have shown six isoenzymes in each group. The third fastest moving isoenzyme was associated with the microsomal fraction (S. L. Allen, 1964). The results have indicated that certain isoenzymes have peak activities during the logarithmic phase of growth and others during the stationary phase.

The effect of adding the non-ionic detergent Triton X-100 to the incubation media on the zymograms of the esterases and acid phosphatases of *T. pyriformis* has been studied (Allen *et al.*, 1965). This detergent seems to be both a differential activator and a differential inhibitor of some of the esterase components.

The acid phosphatases of variety 1 of *T. pyriformis* have been separated into seventeen zones by starch-gel electrophoresis at pH 7·5 (Allen *et al.*, 1963a). All had the same pH optimum and all were inhibited by 10 mM fluoride or 10 mM tartrate. Some of the phosphatases could be characterized by substrate specificity or sensitivity to inhibitors such as *p*-chloromercuro-benzoic acid. Some have been detected only in certain genotypes (Allen *et al.*, 1963a) and the so-called P1 phosphatases were found to be strain specific; those from strain B appeared less heat stable and were activated at a lower pH than those from strain A. It appeared most likely that these P1 phosphatases are controlled by alleles at a single locus or by linked genes (Allen *et al.*, 1963b). F_1 hybrids from so-called strains A and B have been shown to possess a hybrid enzyme, as well as the two parental types. These workers have also been able to describe cell lines with different phenotypes after phenotypic drift; some possessed three and others five components.

5. ALGAE

Some elegant experiments by Keck (1961) have implied that in *Acetabularia* a pre-existing species-specific acid phosphatase could be structurally modified by transplanted nuclear or cytoplasmic factors from another species. Electrophoresis of crude homogenates of *Acicularia schenckii* (*acic*) and *Acetabularia mediterranea* (*med*) gave a single zone of acid phosphatase, the mobility of the enzyme from the latter species being the greater. Interspecies nuclear transplantation, i.e. where *med* nuclei were transplanted to *acic* stalks have shown at first only the *acic*-type isoenzyme; about 3 days after transplantation the *med* form could also be detected. In certain grafts an intermediate form of the enzyme was present. An *acic* nucleus apparently could not change the phosphatase type after transplantation to *med* cytoplasm. It appeared therefore that any *med* component invariably dominated over the *acic* component so that only the *med*-type enzyme was produced (Keck, 1961). These findings are shown in Fig. 53(a). If *acic* and *med* rhizoids were dissected from binucleate plants and maintained in culture, the *acic* rhizoids gave rise to plants which could not be distinguished morphologically from the normal *acic* plants, there was, however, frequent appearance of a *med*-type or intermediate type phosphatase (Fig. 53b). Further studies (Keck and Choules, 1964) confirmed these findings and also indicated that *Acetabularia crenulata* was not subject to interactions leading to alterations in the phosphatase isoenzymes.

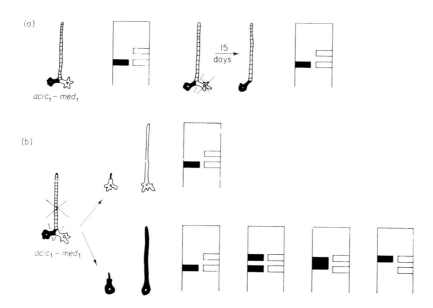

Fig. 53.   Acid phosphatases in algae. Diagrammatic representation of the inability of an *acic*
nucleus to change the phosphatase type of a hybrid cytoplasm (a) and of the persistence of
med traits in a hybrid cytoplasm (b). The latter could be demonstrated by separating the *acic*
and *med* rhizoids from a binucleate plant and maintaining them in culture. The *acic* rhizoids
frequently gave rise to plants which were morphologically identical to normal *acic* plants but
had a *med* type phosphatase. The outlined bands indicate the position of bands obtained
from a mixture of *med* and *acic* (top band) homogenates. The dark bands indicate patterns
associated with grafts, etc. (reproduced with permission from Keck, 1961).

6. BACTERIA

A. ALKALINE PHOSPHATASE

When examined by zone electrophoresis, alkaline phosphatase obtained
from *E. coli* showed a number of enzymatically active bands (Bach *et
al.*, 1961). Mutant forms of this bacterium also produced multi-
molecular forms of alkaline phosphatase which behaved differently
from normal forms on starch-gel electrophoresis. Each isoenzyme, both
in the normal and mutant forms, seemed to be made up of two sub-
units. Those in the mutant form differed from those in the normal form.
Each dimer isoenzyme could be reduced to a mixture of inactive monomers by means of thioglycollate in 6 M urea solution. The active dimer
could be re-synthesized by oxidation in air in the presence of mercaptoethanol. Carrying out this reduction and oxidation procedure on mixtures of alkaline phosphatases obtained from two pseudo-revertants
(Levinthal *et al.*, 1962) has resulted in hybridization with the production

of new active dimer forms of the enzyme. During starch-gel electrophoresis these behave differently from the forms previously known.

Schleisenger and Levinthal (1963) in further studies of the alkaline phosphatase of *E. coli* have also found multiple forms after starch-gel electrophoresis. Proteins antigenically related to the bacterial alkaline phosphatase have been prepared from some phosphatase-negative *E. coli* mutants. These proteins were dissociated into monomers by reduction in urea or by incubation in 50 mM acetate buffer pH 5·0 for 15 minutes at 0°C. Hybrids between these mutants were found to be partially enzymatically active and it appears that the monomers undergo a temperature- and metal-dependent change to form this partially active hybrid. The *E. coli* phosphatase is apparently made up of two identical sub-units. The partially active hybrids are made up from one sub-unit from each mutant strain. Evidence is available that some sub-unit species combine with a greater affinity than others and it is also clear that the specific activity of the partially active hybrid molecules is dependent on the types of monomers combined as hybrids (Fan *et al.*, 1966).

B. OTHER ENZYMES

Starch-gel electrophoresis studies have shown that the electrophoretic mobility of the major reductase activity in the LT-2 (wild type) and *ilva* strain A-8 and D-27 of *Salmonella typhimurium* were different.

The properties of the NAD-dependent triosephosphate dehydrogenase of the phototrophic bacterium *Chromatium* have been shown to differ according to whether the enzyme was grown photolithotrophically with CO_2 as the sole source of carbon and sodium bisulphite as reductant, or photo organotrophically with sodium malate as source of both carbon and reductant (Hudock *et al.*, 1965). The two forms could be interconverted by oxidation or reduction, with corresponding alterations in the Michaelis constants.

Esterases and catalases from four species of rapidly growing mycobacteria, *Mycobacterium phlei*, *M. smegmatis*, *M. fortuitum* and *M. rhodochrous* have been separated into multiple molecular forms by starch-gel electrophoresis using a discontinuous buffer system (Cann and Wilcox, 1965). Species specific esterase patterns could be detected which could be used as means of identification of the mycobacterium.

The heterogeneity of hyaluronate lyase has been examined in various bacteria using chromatography on DEAE cellulose. Strain specific patterns of two or three isoenzymes could be detected in some staphylococci and streptococci (Greiling *et al.*, 1964).

In uninduced lactose-fermenting strains of *E. coli* a single zone of

β-galactosidase could be detected after acrylamide-gel electrophoresis (Appel *et al.*, 1965). On induction with 50 μM isopropyl-β-D-thiogalactoside seven forms of the enzyme were produced. These multiple galactosidases could be separated by molecular sieve techniques indicating that the isoenzymes differed from one another by molecular weight. The single zone of the enzyme had a sedimentation coefficient of 16S. It was thought that the multiple zones were probably made up of differing numbers of sub-units smaller than the 16S component.

CHAPTER IX

Isoenzymes in relation to Clinical Medicine

NUMEROUS observations have now been published in relation to the usefulness of isoenzyme studies in clinical problems. In regard to diagnosis, greatest success has undoubtedly been obtained with serum lactate dehydrogenase and mostly in disease of the heart or liver. Alkaline phosphatase studies are also promising to be quite useful in the differential diagnosis of jaundice. As experience increases, other enzyme systems are coming into diagnostic use. Studies of serum isoenzymes are in fact pointing to the possibility that diagnostic clinical enzymology promises to become a much more accurate procedure.

1. LACTATE DEHYDROGENASE

A. HEART DISEASE

The first reports of elevations of specific fractions of lactate dehydrogenase in serum were given by Hess (1958), Hill (1958) and Vessel and Bearn (1958) but it was not until some time later that the full diagnostic implications of serum LDH isoenzyme patterns were realized. It is now firmly established that the pattern changes in myocardial infarction (Wieme, 1959b; Wróblewski et al., 1960; Latner and Skillen, 1961a; Wróblewski and Gregory, 1961; Van der Helm et al., 1962; Batsakis et al., 1964) and that information of importance to the clinician may be obtained. The type of pattern found in heart muscle extract can be recognized in the serum as early as 6–24 hours after the initial attack of pain and may still be detected 2 or 3 weeks later, even though the level of total serum lactate dehydrogenase has returned to the normal range. Figure 54 shows typical findings in a case of myocardial infarction.

A method of quantitation of infarct size following experimental coronary artery occlusion in dogs has shown that the estimation of LDH-1 in serum correlates more closely with infarct size than any other serum enzyme estimation (Nachlas et al., 1964).

In the study of the serum changes following myocardial infarction, one interesting finding has been the occasional appearance of LDH-5 about 5 or 6 days after the initial attack of pain. This could indicate that some liver necrosis has occurred. It would be tempting to regard this as secondary to venous congestion of that organ but it is difficult to have any precise ideas, since the band does not occur very frequently

146

in ordinary cases of congestive cardiac failure. It must, however, be borne in mind that the slow moving band is also prominent in skeletal muscle and its presence in the serum could possibly be a manifestation of some degree of disuse atrophy.

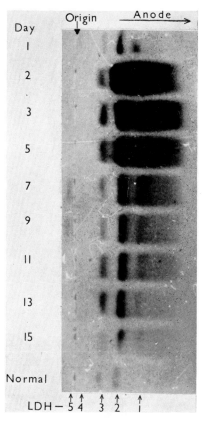

Fig. 54. Serum LDH isoenzymes following myocardial infarction on day 1. A faint "liver" band can be detected on days 7 and 9.

The serum isoenzyme picture is of undoubted importance in deciding when a second myocardial infarction has occurred fairly soon after the original one (Rettenbacher-Daubner and Reider, 1963; Cohen *et al.*, 1964). If this happens, the myocardial type of pattern in the serum suddenly becomes more pronounced and it is quite obvious that additional heart muscle has released its enzyme content into the circulation. In a patient who suffers pain in the chest after the initial incident, it may be difficult to decide which of a variety of causes, including indigestion, could be responsible. Since the electrocardiogram already shows abnormalities, this is occasionally not of much help in deciding whether a second infarction has occurred.

The type of serum pattern following myocardial infarction can be found in other conditions and it is important to realize that it does not necessarily indicate coronary occlusion, although this is usually the case. A similar serum pattern can be observed in patients with acute rheumatic myocarditis, as well as in those with haemolytic or pernicious anaemia.* The electrocardiogram changes obtained in the latter condition may closely resemble those of myocardial infarction and it would be tempting to suggest that the appearance of LDH-1 in the serum of these patients is due to myocardial anoxia. It is now known, however, that in this disease the isoenzyme is mainly derived from marrow tissue (Emerson et al., 1964). After successful treatment with vitamin B_{12} the isoenzyme pattern and the electrocardiogram return quite rapidly to normal. Other conditions which can show patterns closely resembling those of cardiac infarction are generalized carcinomatosis and haemolytic anaemia (Van der Helm et al., 1962).

The isoenzyme pattern in angina is usually normal during the attack (Rettenbacher-Daubner and Reider, 1963; Cohen et al., 1964). Normal patterns have been found not only in anginal attacks but also after pulmonary embolism (Cohen et al., 1964). This is in contrast to the results of others (Rettenbacher-Daubner and Reider, 1963) who thought that lactate dehydrogenase isoenzymes gave false-positive indications of myocardial infarction in some cases of pulmonary embolism or angina pectoris. These observers suggested that repeated electrocardiograms constituted the most reliable method for detection of transmural myocardial infarction, but indicated the importance of isoenzyme studies where there is progressive or recurrent infarction and when electrocardiograms may be difficult to interpret, as in cases of ventricular tachycardia and branch-block. Lactate dehydrogenase isoenzyme patterns are, however, generally considered of use in differentiating between myocardial infarction and pulmonary embolism. The latter condition may be occasionally associated with an increase in serum aminotransferase (SGOT) but the lactate dehydrogenase isoenzyme pattern differs from that seen after myocardial infarction (Van der Helm et al., 1962; Amelung, 1963; Wróblewski, 1963; Mager et al., 1966).

Preston et al. (1965) have reported clinical studies in which serum LDH isoenzymes were separated by cellulose acetate electrophoresis. They found that a pattern similar to that of heart muscle could be

---

* The level of serum lactate dehydrogenase is usually much higher in untreated pernicious anaemia than in myocardial infarction.

detected in patients with myocardial infarction, active rheumatic carditis and acute viral myocarditis; normal patterns were present in patients with angina, pericarditis, heart block with Stokes-Adams syndrome and pulmonary embolism. Serum isoenzyme patterns similar to those found by mixing dilute liver extract with normal serum were seen in patients with congestive heart failure.

Controlled investigation is being undertaken of the serum isoenzyme pattern during prolonged angina (Latner, 1964). The evidence would seem to indicate that some cases have the myocardial pattern and are probably myocardial infarctions and the remainder, forming the vast majority of these patients, either have an infarct which is too small to detect or do not suffer necrosis of heart muscle.

B. LIVER DISEASE

In the serum of patients with jaundice caused by diseases of the liver or biliary system, LDH-5 is more or less constantly present (Hess and Walter, 1960; Latner and Skillen, 1961; Wieme and Van Maercke, 1961; Wróblewski and Gregory, 1961). In hepatocellular disease, however, the band tends to be more prominent and an experienced observer can recognize, with some degree of accuracy, to which particular group a patient belongs. It is interesting to note that LDH-5 also appears in the serum in the pre-icteric phase of infective hepatitis (Latner, 1962) and this finding can be of great diagnostic help. Hepatotoxic drugs can also result in the appearance of this band (Wróblewski and Gregory, 1961; Latner, 1962).

Differentiation between hepatic and haemolytic hyperbilirubinaemia has been achieved by high voltage electrophoresis of serum lactate dehydrogenase on membrane foil (Woerner and Martin, 1961).

Experimental evidence for a rise in LDH-5 resulting from liver necrosis has been obtained by Wieme and Van Maercke (1961) who administered carbon tetrachloride to rats and cynomologus monkeys by intubation. A rise in the isoenzyme could be observed within 24 hours, since the damaged cells released their enzymes into the circulation. The LDH-5 level, however, decreased if the carbon tetrachloride treatment was continued. It would seem, therefore, that continuation of the dose depresses the normal metabolic processes of the liver cells.

Secchi *et al.* (1964) have investigated the problems involved in evaluation of the LDH isoenzyme patterns of human liver needle biopsy specimens. They have shown that it is advisable to use the biopsy specimen as soon as possible after removal; frozen specimens are not as satisfactory as fresh unfrozen material, due to the cold lability of the slower moving iosenzymes. The patterns obtained from fresh needle biopsies and from post-mortem liver are significantly different, possibly due to greater contamination of the post-mortem tissue with blood. No differences in the isoenzyme patterns of needle biopsy specimens of normal and cirrhotic liver have been observed (Secchi *et al.*, 1965).

C. MUSCULAR DYSTROPHY

Studies employing gel electrophoresis have also been undertaken in regard to the distribution of lactate dehydrogenase in skeletal muscles and sera from patients with muscular dystrophy. In the latter an increase in LDH-5 has been reported (Wieme and Herpol, 1962) although this has not been confirmed (Lauryssens *et al.*, 1964). In biopsies from human dystrophic muscle there have been marked differences from normal adult muscle (Dreyfus *et al.*, 1962; Wieme and Lauryssens, 1962; Lauryssens *et al.*, 1964; Lowenthal *et al.*, 1964). LDH-5 which is most prominent in normal adult muscle has been absent or markedly decreased in myopathic muscle. In proven female carriers of Duchenne muscular dystrophy, a reduction in the LDH-5 content of muscle extracts has been found (Emery, 1964).

Pearson *et al.* (1965) have described muscle LDH isoenzyme patterns in two types of X-linked muscular dystrophy; LDH-4 and LDH-5 were greatly reduced in muscles of children with Duchenne dystrophy. In another more benign type of dystrophy extracts of gastrocnemius muscle showed a predominance of LDH-5 and extracts of quadriceps muscle diminished amounts of LDH-4 and LDH-5.

In the muscles of patients with sex-linked muscular dystrophy, the content of SHBD and LDH is lower than in normal muscle but the ratio of the SHBD to LDH activities is much higher. This suggests that the dystrophic muscle contains relatively more of the faster moving LDH isoenzymes and this has been confirmed by acrylamide-gel electrophoresis (Johnston *et al.*, 1965 cited in Wilkinson, 1965a).

Katz and Kalow (1966) have studied the changes in the isoenzyme patterns of lactate, malate and isocitrate dehydrogenases in normal muscle and in myopathies. Only LDH and ICDH showed variations in the diseased muscle but the changes were not specific for any myopathy. The presence of abnormal catheptic activity has been postulated as a possible factor in these alterations in isoenzyme patterns.

It is generally agreed that the diseased muscle is more like foetal than normal adult muscle and investigations of the chicken (Kaplan and Cahn, 1962) and mouse (Lauryssens *et al.*, 1964) confirm the findings on human dystrophic muscle. Investigations of the lactate dehydrogenase isoenzyme patterns of human muscle (Blanchaer and Van Wijhe, 1962; Kar and Pearson, 1963) and other animal muscles (Blanchaer and Van

Wijhe, 1962; Kaplan and Cahn, 1962; Wilson *et al.*, 1963) have shown that the various skeletal muscles within a species differ greatly in iso-enzyme pattern. This has been related to aerobic or anaerobic conditions and it has recently been proposed that the situation in dystrophic muscle is also related to a disparity between anaerobic glycolysis and respiration (Lauryssens *et al.*, 1964). Other workers have noted a reduction or absence of LDH-5 in the skeletal muscles of patients with myotonic dystrophy as well as those with muscular dystrophy (Emery *et al.*, 1965). A reduction in LDH-5 has also been found in one case each of Werdnig-Hoffman disease and dermatomyositis (Emery *et al.*, 1965). Experimental denervation of guinea pig skeletal muscle causes reversion to the foetal type of isoenzyme pattern (Brody, 1965). Following division of one sciatic nerve in rabbits, the resultant denervated muscle showed neural atrophy on histological examination with selective loss of the M type of lactate dehydrogenase (Dawson *et al.*, 1964). It has been suggested that resemblance to the foetal type is a non-specific effect of either myopathic or neurogenic change in skeletal muscle and occurs in hereditary and acquired conditions (Brody, 1965). As red skeletal muscle fibres contain relatively more of the faster moving isoenzymes than do white muscle fibres (Van Wijhe *et al.*, 1964) changes in diseased muscle of the relative proportions of red and white fibres could alter the isoenzyme pattern.

D. DISEASES OF JOINTS

LDH-5 is present in the serum of patients with dermatomyositis and rheumatoid arthritis (Wieme, 1963). It is interesting that in the latter disease LDH-5 is nearly always present, although the total lactate dehydrogenase activity may be normal.

Cohen (1964) has also found increased levels of serum LDH-5 in patients with rheumatoid arthritis. He found similar increase in infective arthritis and gout but normal levels in patients with degenerative joint disease; increased levels of LDH-5 have been found in pathological synovial fluids, as well as increased activity of total lactate dehydro-genase (Vessel *et al.*, 1962b). An investigation of synovial fluid lactate dehydrogenase using electrophoresis and chromatography on both DEAE cellulose and DEAE Sephadex has also shown increased LDH-5 in patients with inflammatory arthropathies (Greiling *et al.*, 1964).

E. RENAL DISEASE

Serum lactate dehydrogenase isoenzymes may be significant in the field of homotransplantation as a biochemical indication of the state of the homograft. Isoenzyme studies of serum alkaline phosphatase may have

G

similar significance, as is described later. An increase in serum LDH-5 has been reported as indicating that homograft rejection will occur (Prout *et al.*, 1964). Increases in LDH-1 and LDH-2, the most prominent isoenzymes in renal tissue, have been detected in the sera of patients who have subsequently rejected their homografts (Latner *et al.*, 1966b). This latter increase, which is found in the last week before death, probably represents necrosis of the renal homograft tissue. During most of the period following the homograft operation increased levels of serum LDH-5 have been detected by most investigators. This may be due to haemodialysis which is always carried out in these cases. Haemodialysis causes increases in serum LDH-5 and Ringoir and Wieme (1965) have suggested that this is due to flushing out of LDH-5 from diseased cells of the renal medulla. These workers have also reported increased serum LDH-5 in many cases of renal tubular disease.

A study employing agar-gel electrophoresis for deriving urinary lactate dehydrogenase isoenzyme patterns in patients with urinary tract disease has shown no consistent elevation of a single isoenzyme, or pattern which could be ascribed to any specific urinary tract disease (Riggins and Kiser, 1964). An increase in urinary LDH-5 has been described in patients with unilateral renal ischaemia (Guttler and Clausen, 1965). Studies of urinary LDH isoenzymes in benign and malignant bladder disease have indicated that leucocytes or erythrocytes or both are the most probable source of increased urinary LDH levels.

F. DISORDERS OF THE NERVOUS SYSTEM

Lactate dehydrogenase isoenzymes have been studied in the grey and white matter of human and sheep brains, with a view to the possible clinical use of isoenzyme patterns in cerebrospinal fluid (Lowenthal *et al.*, 1961a, b, c). Simultaneous changes were found in isoenzyme patterns in both serum and spinal fluid. Investigations of the LDH isoenzymes in cerebrospinal fluid, blood, leucocytes and brain extracts (Van der Helm *et al.*, 1963) have also shown that the pattern is virtually unchanged when lactate dehydrogenase passes the blood-brain barrier. The increased cerebrospinal fluid lactate dehydrogenase level may be due to leakage of the enzyme from leucocytes. Cunningham *et al.* (1965) described up to eighteen zones of LDH after acrylamide disc-electrophoresis of spinal fluid; the additional zones were apparently subfractions of LDH-4 and LDH-5 and probably had little physiological significance. The spinal fluid of patients with organic nervous disease showed significantly increased LDH-2 and LDH-3. It was suggested that this might be of diagnostic significance in destructive nervous lesions. Earlier studies (Lowenthal et al., 1961a) had shown no specific changes in isoenzyme patterns in a number of neurological diseases.

G. PREGNANCY

Studies of human placental extracts have shown that LDH-3 and LDH-4 were most prominent (Meade and Rosalki, 1962). Previous investigators (Wieme and Van Maercke, 1961) had found that 55% of the placental LDH was in LDH-4 and LDH-5, with the other 45% being approximately equally divided between the other three isoenzymes. Hawkins and Whyley (1966) have shown that LDH-3, LDH-4 and LDH-5 were the most prominent isoenzymes in placental extracts and attributed most of the LDH-1 and LDH-2 to contained blood.

Meade and Rosalki (1963) have demonstrated raised levels of LDH-3 and LDH-4 in maternal blood, especially during labour, and have suggested that the increased enzyme activity is of placental origin. Hawkins and Whyley (1966) have indicated that the increased serum LDH_4 and LDH_5 found during labour is derived from uterine muscle. Normal serum LDH isoenzyme patterns have been reported in patients with pre-eclampsia, irrespective of the severity of the toxaemia (Hawkins and Whyley, 1966).

H. MISCELLANEOUS DISORDERS

The serum LDH isoenzyme pattern changes very markedly in patients suffering from acute pancreatitis and comes to resemble closely that which is obtained with an extract of human pancreas. The diagnostic use of this investigation is, however, limited since the pattern can occur in other abdominal emergencies. This is probably due to the associated profound shock, since it has been shown that a very similar pattern can be produced in the serum of dogs suffering from haemorrhagic shock (Vesell and Bearn, 1961).

Kroner and Reunhaver (1964) have examined the serum LDH isoenzyme patterns of guinea-pigs suffering from burns; the pattern was identical with that following carbon tetrachloride poisoning but there was no evidence of enzyme loss from liver, kidney or erythrocytes. It would appear that any increased serum activity was of muscular origin.

The serum pattern of LDH isoenzymes corresponding to that of erythrocytes has been described in patients with pernicious anaemia, homozygous sickle cell anaemia, acquired haemolytic anaemia, lymphoma and the haemolytic stages of leukaemia (Preston et al., 1965). Elliot and Fleming (1965) have presented evidence using SHBD activity to suggest that the red cell precursors of the marrow plasma are the

source of the elevated serum enzyme activity in patients with megaloblastic erythropoiesis secondary to folic acid deficiency.

A report on the serum lactate dehydrogenase isoenzyme patterns of
chimpanzees following impact stress of 54–180 g has shown that LDH-1
and LDH-2 were lower and LDH-4 and LDH-5 higher within 24 hours
after the stress but 11 days later LDH-1 and LDH-2 were found to be
elevated (Hawrylewicz and Blair, 1965). These findings indicate that
some cellular damage in organs such as liver and muscle must occur
during the "blast off" and deceleration periods of space flight.

I. STABILITY OF LACTATE DEHYDROGENASE ISOENZYMES

The question of stability of the lactate dehydrogenase isoenzymes is of
major importance in relation to clinical investigations, as storage of
serum specimens is known to alter the isoenzyme pattern. For many
years total serum lactate dehydrogenase activity was thought to be best
preserved when sera were kept at 0–4°C or frozen at −10°C or −20°C.
However, it has been found by Kreutzer and Fennis (1964) and Zondag
(1963) that storage of sera or tissue extracts in the frozen state preferentially inhibits the slower moving isoenzymes. The former workers
stored sera at room temperature, at 0–4°C and at −10°C and made
sequential studies of the lactate dehydrogenase isoenzyme patterns.
LDH-1 was stable for 1 month at all three temperatures; LDH-2,
LDH-3, LDH-4 and LDH-5 showed no alteration for 10 days at room
temperature. At −10°C there was a complete loss in LDH-4 activity
within 2 days while LDH-5 fell off within 8 days. LDH-2 and LDH-3
lost half their activity within 8 days and were inactive within 1 month.
Zondag (1963) showed that tissue extracts stored at −20°C lost LDH-5
activity first, but that sera and extracts with added NAD (10 mg/ml)
or glutathione (10 mg/ml) were much more stable when frozen. In
general, lactate dehydrogenase isoenzyme patterns should be determined only on reasonably fresh samples and storage for up to a few
days should be at room temperature. For longer periods it may be
necessary to store at 4°C to avoid contamination with organisms as far
as possible.

J. SIMPLE LABORATORY TESTS FOR ISOENZYMES

(i) *Heat stability*. Following the demonstration of different heat stabilities
for the five lactate dehydrogenase isoenzymes (Plagemann *et al.*, 1960b),
a simple test was developed which differentiated between the faster
moving and slower moving isoenzymes (Wróblewski and Gregory,
1961). The test involved dividing a serum sample into three portions,
heating one portion to 65°C for 30 minutes, to obtain LDH-1 activity;

leaving the second portion at room temperature to give total activity; and heating the third portion to 57°C for 30 minutes, to obtain LDH-5 activity by difference from the total. Each portion was now assayed for LDH in the usual way.

Many workers have used heat stability techniques for determining which serum isoenzyme fractions are increased in disease states. They have proved most useful as an aid to the diagnosis of myocardial infarction (Dubach, 1961; Stranjord and Clayson, 1961; Stranjord et al., 1962; Wust et al., 1962; Bell, 1963; Nutter et al., 1966). Paunier and Rotthauwe (1963) have used heat stability to identify the isoenzyme patterns in a number of heart, muscle and liver diseases and other disorders.

A "heat stability index" (Latner and Skillen, 1963) of serum lactate dehydrogenase has proved to be a useful diagnostic tool in myocardial infarction. This index is the ratio of the LDH activity of serum heated at 60°C for 1 hour to the activity of unheated serum.

(ii) *Action of organic solvents*. Addition of acetone to serum inactivates all isoenzymes except LDH-1 and this effect has been used in a simple test for this isoenzyme in serum (Latner and Turner, 1963). If 0·5 ml serum is mixed with 0·5 ml water and 0·5 ml acetone added to the mixture, the resultant LDH activity can be attributed solely to LDH-1. As this is the isoenzyme most prominent in cardiac muscle, this technique has clinical application in the diagnosis of myocardial infarction. Patients with untreated pernicious anaemia and certain malignant disorders also show high levels of LDH-1 (Latner and Turner, 1963). Other workers have confirmed the usefulness of this technique in the diagnosis of myocardial infarction and indicated that it is more sensitive than heat-stability tests (Englisova et al., 1964).

Chloroform has also been shown to inhibit all except the fastest moving serum LDH isoenzyme (Warburton et al., 1963). If diluted serum is shaken with chloroform, the enzyme activity remaining is a good measure of the amount of the serum enzyme which is of cardiac origin. Evidence that the total lactate dehydrogenase activity of the red cell is inhibited by chloroform to a greater extent than is that of serum can be attributed to differences in absolute protein concentration (Warburton, 1965).

(iii) *Oxalate and urea*. The effects of oxalate or urea on serum LDH activity have been discussed in Chapter V. It has been shown (Emerson and Wilkinson, 1965) that more than 68% oxalate inhibition and less than 45% urea inhibition of serum lactate dehydrogenase with pyruvate as substrate is indicative of myocardial infarction and that less than 50% oxalate inhibition and more than 62% urea inhibition under the same conditions is indicative of hepatic disease. Endogenous urea is unlikely to have any effect under the conditions of the test.

Other workers have used inhibition with 1·5 M urea as a means of detecting whether elevated serum lactate dehydrogenase levels are of hepatic or cardiac origin (Kontinnen and Lindy, 1965) and inhibition with 2·6 M urea has also been suggested as a means of detecting the heart isoenzymes in serum (Hardy, 1965).

(iv) α-*hydroxy-butyrate dehydrogenase*. By comparing the activity of a serum with oxobutyrate as a substrate to the activity with pyruvate as a substrate, it is possible to obtain an indication of the relative amounts of LDH-1 and LDH-5 present. The more closely the index approaches unity, the more LDH-1 is present and vice-versa (Rosalki and Wilkinson, 1960). Similar findings have been obtained using the reaction in a reverse direction (Plummer *et al.*, 1963). The value of serum hydroxybutyrate dehydrogenase (SHBD) estimations is mainly in the diagnosis of myocardial infarction, where elevation of the enzyme can be detected 12 hours after the attack and the highest level is reached in 48–72 hours. The abnormal levels can often be detected 2–3 weeks after the initial infarction (Elliot and Wilkinson, 1961; Kontinnen, 1961; Pagliaro and Notarbartolo, 1961; 1962; Elliot *et al.*, 1962; Hansson *et al.*, 1962; Kontinnen and Halonen, 1962; Rosalki, 1963; Wilkinson and Rosalki, 1963; Preston *et al.*, 1964; Rosalki and Wilkinson, 1964).

Wilkinson (1965b) has reviewed the diagnostic use of this test and reported that normal values are usually found in pulmonary infarction, acute cholecystitis, hiatus hernia, congestive cardiac failure, pericarditis, pleurisy and rheumatic valvular disease of the heart. Special attention has been drawn to the values in patients with angina of effort; of sixty-seven cases, all except eight cases with borderline SHBD levels had normal values. Raised SHBD levels have, however, been reported in myocarditis (Elliot *et al.*, 1962) and after cardiac surgery (Pyorola *et al.*, 1963).

Raised levels can be detected in hepatocellular disease but in these cases the ratio of SHBD to LDH activities has been much lower than that found in normal healthy individuals (Elliot and Wilkinson, 1963). Low ratios have also been found in patients with cirrhosis or obstructive jaundice (Wilkinson, 1965b).

High serum SHBD levels are present in many patients with malignant diseases. The SHBD/LDH ratio in these patients has usually been normal, unless there was evidence of hepatic metastases (Wilkinson, 1965b).

2. Aspartate aminotransferase

The finding of specific mitochondrial and cytoplasmic components raises the possibility that in less severe diseases the latter might appear in the circulation. In severe diseases, involving the death of many cells, the mitochondrial entity might appear. It has been shown that the latter component appeared in the blood of a patient with severe carbon tetrachloride poisoning, as well as in another patient with a massive myocardial infarction (Boyde and Latner, 1962).

There are indications from animal experiments of possible clinical applications in the field of haematology. Mature rabbit erythrocytes contain only the anionic or cytoplasmic isoenzyme of aspartate amino-transferase (Nisselbaum and Bodansky, 1965). Reticulocytosis induced by massive bleeding or acetylphenylhydrazine causes an increase in erythrocyte enzyme activity and the appearance of the cationic or mitochondrial enzyme. In normal human adult red cells only the cyto-plasmic enzyme is present, whereas in the erythrocytes of patients with reticulocytosis the mitochondrial enzyme can be detected (Mannucci and Dioguardi, 1966).

Starch-gel electrophoresis of aspartate aminotransferase in serum and tissues of patients with myopathies has shown that the mitochondrial enzyme cannot be detected in the serum and it has been concluded that the mitochondria are not undergoing breakdown in myopathies (Kar and Pearson, 1964).

3. Creatine phosphotransferase

No alteration in the creatine phosphotransferase isoenzyme patterns of human muscle extracts have been detected in patients suffering from common myopathies (Kar and Pearson, 1965). As a result of experi-mental denervation of the gemellus muscle in the rabbit, Schapira (1966) has found that the muscle isoenzyme pattern reverts to the type found in foetal muscle.

Van der Veen and Willebrands (1966) have detected one or two isoenzymes in sera from patients with muscular dystrophy. The same two isoenzymes occur after myocardial infarction. A possible diagnostic use of serum creatine phosphotransferase isoenzymes arose from the observations of Acheson et al. (1965), who reported significant increases in the serum levels of this enzyme following cerebral infarction. This has led to the investigations of Kalbag et al. (1966), who separated the enzyme into two fractions using adsorption-elution on DEAE-Sephadex. Approximately 90% of the isoenzyme derived from brain occurred in one fraction and approximately 90% of the muscle isoenzyme occurred in the other. Using this technique these workers have made the interesting finding that the raised serum creatine phosphotransferase following brain damage due to strokes is derived from skeletal muscle rather than from brain.

4. ALKALINE PHOSPHATASE

A. LIVER AND BONE DISEASE

In sera with elevated serum alkaline phosphatase levels obtained from patients with liver disease, it has been shown that the main bands of activity move to different positions from those found with sera from patients with bone disease (Hodson *et al.*, 1962). In Fig. 55 it can be seen that each of the sera from the patients with disease affecting the liver had a main band of activity which moved more rapidly than that of the patients with bone disease, viz. parathyroid osteitis, Paget's disease and osteomalacia. It has been suggested that in each of these

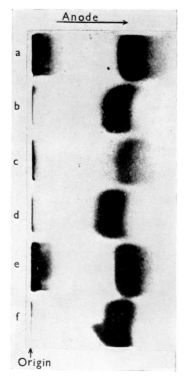

FIG. 55.   Serum alkaline phosphatase isoenzyme patterns in bone and liver diseases. a, biliary obstruction; b, parathyroid osteitis; c, hepatocellular disease; d, Paget's disease; e, biliary obstruction; f, osteomalacia.

two disease groups the serum phosphatases partially purified by starch-gel electrophoresis have the same "K_m" values as those of the corresponding tissue extracts (Moss et al., 1961a; 1961b).

When jaundice is due to obstruction of the biliary tract, it has been possible by starch-gel electrophoresis of the serum to demonstrate a band or couplet of alkaline phosphatase activity in the β-lipoprotein region (Hodson et al., 1962). It would seem that this observation has diagnostic application. A similar conclusion has been reached by other workers (Chiandussi et al., 1962). A group of fifty-four consecutive cases of jaundice of definitely proven aetiology (Latner, 1965) has been examined from this point of view. The results are shown in Table 12. It

TABLE 12

Incidence of slow moving serum alkaline phosphatases in patients
with different types of jaundice

Diagnosis	No. of cases	No. showing slow moving bands	% showing slow moving bands
Extra hepatic obstruction	31	29	93·5
Hepatocellular	7	1	14
Cirrhosis (portal)	11	9	82
Haemolytic	5	0	—

will be seen that there is an undoubted association between the frequency of occurrence of activity in the β-lipoprotein region and obstruction of the biliary tree. Presumably, in hepatic cirrhosis, such obstruction is due to fibrous tissue. The β-lipoprotein moiety has been attributed to regurgitation of alkaline phosphatase from bile (Chiandussi et al., 1962).

In the elderly jaundiced patient with an enlarged liver and a raised level of serum alkaline phosphatase, starch-gel electrophoresis can prove successful in differentiating metastasizing carcinoma from primary disease of the liver (Latner, 1965). In the former case the bone isoenzyme is often found and not that from the liver.

Other investigators (Taswell and Jeffers, 1963) have demonstrated eight separate zones of serum alkaline phosphatase activity which it is suggested can be separated into three distinct patterns; one of which is associated with obstructive, metastatic or infiltrative hepatobiliary disease; a second with hepatic parenchymal disease; and a third with bone disorders associated with increased osteoblastic activity. Patterns of activity which are clinically significant can also be obtained after agar-gel electrophoresis (Haije and De Jong, 1963). Here the faster moving main liver band is separated into two portions. Studies have also been made after cellulose acetate strip electrophoresis (Korner, 1962) and alkaline phosphatase activity has been found in positions corresponding to α_1-, α_2- and β-globulins and albumin. In bone disease there is mainly an increase in the β-globulin fraction, whereas in liver disorder the main increases are in the α_2-globulin fraction.

Using a combination of techniques such as starch-gel electrophoresis and column chromatography on DEAE Sephadex, Kotzaurek and Schobel (1963) failed to differentiate between the serum alkaline phosphatase isoenzyme patterns in liver and bone disease, but were able to demonstrate some additional phosphatase isoenzymes in patients with cancer of the head of the pancreas or obstructive jaundice.

In the presence of 20% ethanol most of the phosphatase activity in the sera of patients with bone disease is precipitated, whereas in liver disease the enzyme activity remains in the supernatant (Peacock et al., 1963).

A study of the L-phenylalanine sensitive alkaline phosphatase in sera of normal subjects and patients with cancer or cirrhosis has shown that those with cancer have a lowered proportion of intestinal alkaline phosphatase, whereas in ten out of thirty-three patients with cirrhosis the proportion of intestinal phosphatase was higher than normal (Fishman et al., 1965; Kreischer et al., 1965). There was some evidence to suggest that there was an association between elevated serum intestinal phosphatase and portal hypertension. Newton (1966) has also identified a serum alkaline phosphatase isoenzyme of intestinal origin by electrophoretic mobility and phenylalanine inhibition. This serum isoenzyme could be detected in 20% of healthy individuals and in patients with a variety of diseases. Marked elevations have been detected in patients with portal hypertension and cirrhosis. In hypophosphatasia it has been shown that the decrease in the total serum alkaline phosphatase is not due to a reduction in the activity of this particular isoenzyme.

B. RENAL DISEASE

It has recently been shown (Latner, 1965; Latner et al., 1966b) that the isoenzyme pattern of kidney tissue appears in the serum when a homograft is being rejected. This might well develop into a useful method for detecting complete or partial rejection of the grafted organ.

It has also recently been reported (Moss, 1964; Butterworth et al., 1965) that a fast moving isoenzyme of alkaline phosphatase can appear in human urine. Greatest amounts of this seem to occur in relation to the onset of diuresis in patients who have suffered acute tubular necrosis.

The output apparently occurs in sharp peaks. The isoenzyme also appears in association with "nephritic" exacerbations occurring in patients with the nephrotic syndrome, even though the exacerbations may not necessarily be clinically obvious. The exact source of this isoenzyme has not been determined and it has a smaller molecular weight than those present in kidney extracts. Urines from patients having high alkaline phosphatase levels in their blood not infrequently give rise to an alkaline phosphatase isoenzyme pattern very similar to that in the circulation. A rejected kidney homograft can give rise to the appropriate pattern in the urine.

C. MALABSORPTION STATES

A recent report has described serum alkaline phosphatase isoenzymes after partial gastrectomy and in other malabsorption states (Yong, 1966). The serum alkaline phosphatase isoenzyme thought to be derived from the intestine has been detected in six out of fourteen patients after partial gastrectomy and six out of twenty patients with steatorrhoea due to other causes. The incidence in each case is, however, no greater than expected from observations on normal individuals.

D. PREGNANCY

The changes of serum alkaline phosphatase isoenzymes in pregnancy have already been discussed (Chapter IV).*

5. ACID PHOSPHATASE

The first few reports (de Grouchy, 1958; Estborn, 1959; 1961) indicated the presence of only a single zone of enzyme activity after starch-gel electrophoresis of sera from normal individuals and patients with prostatic carcinoma. Dubbs et al. (1960), however, found three isoenzymes of acid phosphatase in normal sera. Five zones have been reported in the sera of patients with Gaucher's disease (Grundig et al., 1965). Four zones of acid phosphatase have also been described in the sera of some patients with prostatic carcinoma, whereas in normal individuals and patients with primary hyperthyroidism or osteoporosis, three zones were obtained. The substrate specificities and sensitivity to differential inhibitors of the serum enzymes in these diseases have also been described (Grundig et al., 1965). Goldberg et al. (1966) have examined the serum acid phosphatase isoenzyme patterns in patients with Gaucher's disease, prostatic carcinoma and multiple myeloma. In normal serum and in sera from patients with prostatic carcinoma or

* The heat stable isoenzyme is now of importance as an index of placental function. See Messer, R. H. (1967); Appendix p. 238.

myeloma they found a single zone of enzyme activity but in patients with Gaucher's disease, at least two zones were present; as many as four additional zones being present in some of these sera.

Using immunoelectrophoresis, Shulman et al. (1964) found two isoenzymes of acid phosphatase in prostatic extracts; no staining of the precipitin lines could be detected in normal sera but there was some staining in certain patients with carcinoma of the prostate. Further observations have demonstrated four or five forms of acid phosphatase in the tissue and secretions of the prostate (Shulman and Ferber, 1966).

6. Arylamidase

The clinical significance of isoenzymes of arylamidase (leucine amino-peptidase) has been investigated by several groups of workers using starch-gel electrophoresis. In certain pathological sera, especially from patients with pancreatitis, up to eight zones of activity could be detected (Dubbs et al., 1961; Schobel and Wewalka, 1962).

Abnormal arylamidase isoenzyme patterns were detected in most liver disorders and the use of such isoenzyme patterns has been recommended for differentiating "cholangiolitic" hepatitis from acute paren-chymal hepatitis, following the clinical course of hepatitis and defining the degree of biliary cirrhosis (Kowlessar et al., 1960). Specific serum patterns have been found in intrahepatic cholestasis, cholangitis and acute pancreatic disorders (Schobel and Wewalka, 1962).

The possible usefulness of arylamidase isoenzymes in the differentia-tion of acute pancreatitis from other acute abdominal emergencies has been reported (Latner, 1963). It has been found that, whereas normal serum contains only one band in the fast α_2-globulin region, sera from patients with acute pancreatitis contain at least three zones; one in the same position as the normal serum, another between the slow α_2- and β-globulins and a third close to the insertion slot. Similar patterns have been observed in some cases of carcinoma of the gastro-intestinal tract. In liver disorders only the fastest moving zone was usually present.

Investigations of arylamidase isoenzymes by cellulose acetate electro-phoresis of pathological sera have shown that the serum isoenzyme patterns were probably of only limited clinical significance (Meade and Rosalki, 1964).

7. Esterases

It has already been mentioned that in normal human serum, cholin-esterase appears as a doublet after starch-gel electrophoresis in a dis-continuous system. Both these components are normally inhibited by 10^{-4} M eserine. In patients with succinylcholine apnoea one of the components is not so inhibited (Latner, 1962). Serum from a patient with a genetically determined absence of pseudocholinesterase has shown no activity after starch-gel electrophoresis, whereas ten fractions have been observed in a concentrated serum pseudocholinesterase preparation; under the same conditions fresh human serum shows only one single component (Harris *et al.*, 1962). The genetic implications of the findings with this enzyme have already been discussed (Chapter VII).

In patients with cirrhosis there is a marked reduction of esterase activity in serum and hepatic tissue; some of the zones of aromatic esterase normally present cannot be detected in the serum (Secchi and Dioguardi, 1965).

Barron *et al.* (1963) have been able to show abnormalities in the brain esterase isoenzyme pattern in multiple sclerosis. There appeared to be a variable loss in activity of several of the isoenzymes and a hydrolase of α-naphthyl propionate characteristic of white matter could not be found in multiple sclerosis plaques. Further observations on the brain esterase and phosphatase isoenzyme patterns in multiple sclerosis and a possible allied disorder, Schilder's disease, have been reported (Barron and Bernsohn, 1965). Quantitative and qualitative alterations in the non-specific esterases have been found in these diseases. As well as some non-specific abnormalities, a loss of the enzymes which preferentially hydrolyse C_3—C_5 naphthol esters has been detected in demyelinated white matter. A loss of one or more of the multiple molecular forms of cholinesterase has also been described in the demyelinated tissue (Barron and Bernsohn, 1965).

Aberrant patterns of esterase and cholinesterases have been detected in tissues of some mice with neuromuscular mutations (Meier *et al.*, 1962).

8. ISOENZYMES IN CANCER

A. LACTATE DEHYDROGENASE

It has been known for a long time that serum lactate dehydrogenase is frequently elevated in neoplastic disease (Hill and Levi, 1954). The finding of an abnormal elevation in the serum is by no means constant nor can it be regarded as specific. The level in leukaemia seems, however, closely to parallel the clinical status (Bierman *et al.*, 1957). In mice with transplanted and induced tumours the serum lactate dehydrogenase activity rises only if the tumours are malignant (Hsieh *et al.*, 1955). It is known that malignant tissues tend to contain a somewhat higher level of lactate dehydrogenase than the corresponding normal tissues (Meister, 1950). It was, therefore, only to be expected that lactate dehydrogenase isoenzyme studies should be applied in the investigation of cancer.

It has been shown that malignant tumours contain mainly LDH-3, LDH-4 and LDH-5, irrespective of the tissue from which the tumour arose (Pfleiderer and Wachsmuth, 1961; Bar *et al.*, 1963; Richterich *et al.*, 1963; Goldman *et al.*, 1964; Poznanska-Linde *et al.*, 1966). LDH isoenzyme patterns in which these slower moving isoenzymes are the most prominent have also been described in tumours of the cervix (Turner, 1964; Latner *et al.*, 1966c), the breast (Richterich and Burger, 1963; Barnett and Gibson, 1964) and the brain (Gerhardt *et al.*, 1963). An increase of LDH-5 in all types of malignant tumours of the central nervous system has been described. This is in contrast to an increase in the faster moving isoenzymes in the relatively benign gliomas, especially oligodendrogliomas (Gerhardt *et al.*, 1963). Examination of the lactate dehydrogenase isoenzyme patterns of human prostatic carcinoma and of prostates showing benign hyperplasia has shown that in the tumour there is a relative increase in LDH-5 (Denis *et al.*, 1962; 1963). In relation to this finding it must also be noted that increase in LDH-5 has been detected in extracts of canine prostatic tissue of castrated mongrels after administration of testosterone (Denis and Prout, 1962). LDH-3 has been reported as being the most prominent isoenzyme in two studies of lactate dehydrogenase in tumours (Starkweather and Schoch, 1962; Bottomley *et al.*, 1964).

Baume *et al.* (1966) have described the LDH isoenzyme patterns in extracts of fundic and pyloric mucosa. LDH-1, LDH-2 and LDH-3 have been found to be the more prominent isoenzymes in fundic mucosa and LDH-2, LDH-3 and LDH-4, the more prominent in pyloric mucosa. The pattern of gastric tumour tissue resembles that of pyloric mucosa.

In effusions associated with malignancy the slowest moving isoenzymes have been shown to be prominent (Richterich *et al.*, 1962; Richterich and Burger, 1963) and in cases of metastatic carcinoma of the brain a similar pattern is found in the cerebrospinal fluid (Van der Helm *et al.*, 1963). These abnormal patterns in cerebrospinal fluid may be of diagnostic significance in regard to differentiation from other cerebral lesions. The differentiation of malignant and benign pleural effusions is complicated by the demonstration that necrotic tuberculosis produces a "malignant" and productive tuberculosis a "benign" pattern (Burgi *et al.*, 1964).

Studies on the heterogeneity of lactate dehydrogenase in the blood cells and sera of leukaemia patients have shown that in the myeloblastic form LDH-2 is the most prominent isoenzyme and that in myeloid and megaloblastic leukaemia LDH-2 and LDH-3 are increased (Dioguardi *et al.*, 1962; Woerner and Martin, 1962; Yakulis *et al.*, 1962). Examination of the lactate dehydrogenase isoenzyme patterns in normal and leukaemic lymphocytes has indicated that in normal lymphocytes LDH-2 is predominant, whereas in mature and small leukaemic lymphocytes LDH-3 is the most prominent fraction (Dioguardi and Agostini, 1965). It has been shown by other workers that distribution of the isoenzymes in leukaemic cells seems to be the reverse of that found with other malignant tissues; in acute leukaemias, for example, there is usually a preponderance of the faster moving lactate dehydrogenase isoenzymes (Latner and Turner, unpublished observations).

Similarities to the isoenzyme patterns of tumour tissues can be recognized in tissue cultures of human and other animal cell lines, if these are kept for a prolonged period. The cultures develop the pattern characterized by a preponderance of the slower moving isoenzymes (Philip and Vesell, 1962; Vesell et al., 1962; Nitowsky and Soderman, 1964). A similar pattern can be detected in cultures of HeLa and HEP 2 cells which have been going for many years. These are, of course, cultures of cancer cells from the beginning.

Infection of tissue cultures of monkey kidney cells or human thyroid tissue by adenovirus 12, which is known to be oncogenic, apparently causes the culture to develop the malignancy type of pattern well before control uninfected cultures (Latner et al., 1965). It is interesting to note that the total lactate dehydrogenase is also increased. Cultures infected with poliovirus, however, differ in no way from the controls with regard to lactate dehydrogenase.

Infection of mice with the Riley lactate dehydrogenase elevating agent results in an increase of the slowest migrating isoenzyme in the plasma. The plasma of infected tumour-bearing mice contains more of this isoenzyme than does that of infected mice without tumours (Plagemann et al., 1963). No significant differences have, however, been detected in tissues of mice infected with this virus when compared with normal uninfected tissues (Warnock, 1964).

Examination of the serum lactate dehydrogenase isoenzymes in rats bearing experimental tumours and those suffering from acute liver disease such as CCl_4 poisoning or viral hepatitis, has shown that in the tumour-bearing animals there was an absolute or relative increase in LDH-3, whereas in the animals with acute liver disease there was an increase in LDH-5 (Cenciotti and Mariotti, 1964). Johnson and Kampschmidt (1965) have shown that after administration of carcinogens to rats the LDH isoenzyme pattern of liver extracts shows the characteristic changes associated with carcinogenesis within 12 weeks and that the LDH isoenzyme pattern of the rat plasma is similar to that of the tumour tissue.

Serum lactate dehydrogenase isoenzyme patterns have been examined in patients with malignant tumours of the testis or ovary using agar-gel electrophoresis (Zondag, 1965). The pattern usually found in such malignancies was a relative increase in LDH-2, LDH-3 and LDH-4 even though the total levels were normal. In patients with seminoma, teratoma testis and ovarian dysgerminoma the serum pattern resembled that found after myocardial infarction, viz. an increase in LDH-1 and LDH-2. An elevation of serum LDH-4 and LDH-5 has also been detected in some cases of prostatic carcinoma (Denis *et al.*, 1962).

Starkweather *et al.* (1966a) have made a detailed study of agar-gel electrophoresis of serum lactate dehydrogenase with the object of providing a reproducible method for clinical studies. They have shown that although there is a marked diurnal variation in total serum LDH the isoenzyme pattern does not change. This study has been extended to the changing serum LDH isoenzyme patterns during therapy for lung cancer (Starkweather *et al.*, 1966b). When the total serum LDH activity rose rapidly and the isoenzyme pattern was similar to that of the tumour, the patient's clinical condition deteriorated. Non-specific changes in the pattern usually indicated that the disease was nearing its terminal state.

In patients with renal tumours, Macalalag and Prout (1964) have shown that urinary LDH-4 and LDH-5 levels have been markedly elevated when compared with normal subjects and patients with simple renal cysts. From examination of the lactate dehydrogenase isoenzyme pattern of homogenates of normal and cancerous renal tissue, they concluded that the tumour-bearing kidney was the source of the elevated urinary enzyme. Other workers, however, found that the isoenzyme pattern of urine from patients with bladder malignancies could not be distinguished from that from patients without malignancies (Gelderman *et al.*, 1965). They suggested that the elevated urinary lactate dehydrogenase activity of patients with malignant and non-malignant bladder disease was derived from leukocytes and erythrocytes in the urine rather than from the bladder epithelium.

Sometimes unusual isoenzyme bands have been found in cancer tissues or sera from patients suffering from cancer (Latner, 1964). It could be more than a coincidence that a family containing a newly described genetic variant also showed a remarkably high incidence of cancer (Latner, 1964). Lactate dehydrogenase variants have also been detected in a patient with chronic lymphocytic leukaemia and another with lymphoblastic sarcoma (Vesell, 1965d). Agar-gel electrophoresis has revealed the presence of an additional LDH-2 isoenzyme in extracts of tumour tissue from two patients with brain tumours (Soetens *et al.*, 1965).

It has been shown that there was sequential inhibition of the individual lactate dehydrogenase isoenzymes of normal brain and of malignant tumours with increasing concentrations of stilboestrol diphosphate. The sequence was in the direction from the slowest moving isoenzyme towards the fastest moving (Clausen and Gerhardt, 1963). It has, therefore, been suggested that the known cytostatic effect of this substance in the treatment of cancer might be due to the blocking effect on the slower moving lactate dehydrogenase isoenzymes so common in malignancy.

B. GLUCOSE-6-PHOSPHATE DEHYDROGENASE AND PHOSPHOGLUCONATE DEHYDROGENASE

Using starch-gel electrophoresis, the sex-linked electrophoretic variants of glucose-6-phosphate dehydrogenase have been examined in a number of women with leiomyomas (Linder and Gartler, 1965b). Samples of normal myometrium showed both variants, whereas all leiomyomas contained a single zone of either type. Tumours containing each type only were found in all the uteri examined, which suggests that the tumours arise from single cells. It must be emphasized, however, that these tumours are not cancers but have been included in this section for convenience.

An interesting phenomenon has been demonstrated in the writers' laboratory (Latner and Turner, unpublished observations) in regard to the distribution of phosphogluconate dehydrogenase after starch-gel electrophoresis of extracts from cervical carcinomata. Although each malignant tumour showed only one band, the bands appeared at different distances from the origin. This might possibly have been an indication of gene mutation effects within tumours. On the other hand, these variations in mobility could represent indications of aging of the individual cells in the cell population in question. Variations with age in the mobilities of glucose-6-phosphate dehydrogenase and aspartate aminotransferase have been demonstrated (Walter *et al.*, 1965). The electrophoretic mobility of the enzymes from older cells appeared to be greater than that of young cells. It appears, however, that the different mobilities of phosphogluconate dehydrogenase in cancers of the cervix could be explained in another way. Addition of excess NADP to the tumour extracts abolished the inequalities of mobility. It would appear that cancer tissues could contain more of the coenzyme than normal tissues but that the content of the coenzyme varied from tumour to tumour.

H

C. ALKALINE PHOSPHATASE

Starch-gel electrophoresis of normal human leucocyte alkaline phosphatase has demonstrated three electrophoretic variants of the enzyme (Robinson *et al.*, 1965). Butanol extracts of some leukaemic leucocytes gave different patterns from similar extracts of normal leucocytes. In one case of untreated leukaemia only a single zone of phosphatase could be detected and in a number of patients with chronic leukaemia treated with 6-mercaptopurine three zones were found each with different mobilities from the normal leucocyte alkaline phosphatases. Some doubt as to whether these abnormal patterns were due to malignancy arises from the later findings of the same research group (Robinson *et al.*, 1966). Abnormal patterns, some quantitative and some qualitative, have been described in leucocyte extracts of patients with paroxysmal nocturnal haemoglobinuria, pregnant women, and two patients undergoing cortisone therapy; one following castration and hypophysectomy for breast carcinoma and the other with rheumatoid arthritis. Trubowitz and Miller (1966) have also reported that the mobility of the fastest moving zone of alkaline phosphatase in extracts of leucocytes was reduced in patients with polycythemia rubra vera.

D. ASPARTATE AMINOTRANSFERASE

The distribution of aspartate aminotransferase isoenzymes in the homogenate and four sub-cellular fractions of normal embryonic and regenerating liver and in a series of transplantable hepatomas has indicated that there was a correlation between the level of the supernatant enzyme and proliferative activity (Sheid *et al.*, 1965).

E. CHROMATOGRAPHIC STUDIES OF ENZYMES IN RELATION TO CANCER

Chromatography on columns of DEAE cellulose has been used to study enzyme heterogeneity in normal and tumour tissues (Angeletti *et al.*, 1960a; 1960b; Moore and Lee, 1960; Moore and Angeletti, 1961; Bar *et al.*, 1963). Lactate malate and glyceraldehyde phosphate dehydrogenases, acid phosphatase, hexokinase, aldolase and aminotransferases have been among the enzymes studied (Moore and Angeletti, 1961). The isoenzyme patterns in the tumour tissues in each case resembled those of normal tissues but differed in the relative proportions of the different fractions. It has been shown that acid phosphatase exists in multiple forms in organs from the rat, mouse, rabbit and chicken and in general the brain and liver enzymes appeared identical in chromatographic and enzymatic properties. The acid phosphatase isoenzyme pattern of normal rat liver was different from that of rat hepatoma and that of normal epithelium differed from that of a squamous cell carcinoma. No enzymes with abnormal chromatographic properties have been detected (Moore and Angeletti, 1961). Similar findings in relation to multiple forms of lactate, malate and glyceraldehyde phosphate dehydrogenases, enolase, aldolase and pyruvate kinase have been reported with the observation that the changes in the lactate dehydrogenase isoenzyme pattern in tumour tissues were the most striking (Bar *et al.*, 1963).

Chromatography has been used to study the malate and iso-citrate dehydrogenases and aspartate aminotransferases of normal and pathological human and animal livers (Reith *et al.*, 1964).

CHAPTER X

Technical Methods

1. SEPARATION TECHNIQUES

A. AGAR-GEL ELECTROPHORESIS

THE USE of agar as a medium for zone electrophoresis was introduced by Gordon *et al.* (1949). The more refined and elegant immunoelectro-phoresis technique (Grabar and Williams, 1953) led to considerable interest in the use of agar as an electrophoresis medium. Wieme (1959a; 1959b) was the first to elaborate the use of agar-gel electrophoresis in enzyme studies and developed his method of enzymoelectrophoresis which has had numerous applications.

(i) *Preparation of Gel.* A 1% solution of agar (Ionagar No. 2, Oxoid Ltd. or Difco Special Agar-Noble) is made in barbitone buffer (μ 0·01; pH 8·4) by heating to boiling point for a few minutes until the solution is clear. This is most easily done in a boiling water bath. The warm agar solution is filtered through glass wool and a little poured into a shallow glass or Perspex dish. The agar is allowed to set forming a flat surface. Glass lantern plates (3·25 × 3·25 in) or microscope slides (3 × 1 in) are placed on the gel surface and more warm agar poured over the glass plates until the agar layer on top of the plates is about 2 mm thick. Air bubbles are removed with a probe or Pasteur pipette and the gel left to set for 1–2 hours at room temperature. Some workers make up stock agar-gel in water at 2, 3 or 4% and then remelt and dilute with barbitone buffer before use to give a final concentration of 1% agar.

(ii) *Sample Introduction.* The glass microscope slides or plates are cut out from the agar and small cuts about 0·5–1·5 cm in length are made in the gel with a razor blade. The cuts are made at approximately one third of the length from one end of the slide or plate. A small piece of rigid filter paper is inserted into the cut, care being taken not to touch the bottom of the slit. After removing the paper, the protein sample may be introduced into the slot with a Pasteur pipette. Very small tissue fragments may be introduced directly into the groove without homogenization (Wieme, 1959b). The agar plates are then placed in an electrophoresis tank as shown in Fig. 56. The agar plate rests face downwards on the agar blocks which are of identical composition to the plate. The use of agar blocks as bridges between the electrode

170

compartment and the electrophoresis plate helps to prevent any disturbances due to electroendosmosis. A voltage gradient is applied for 25 minutes so that a current of up to 15 mA is carried by each standard 1 inch wide microscope slide. The procedure may be carried out at 4°C or the tank may be filled with petroleum ether, which cools the electrophoresis plate by evaporation (Wieme, 1959a). After each electrophoresis the polarity of the electrodes is reversed so that the agar bridge blocks can be used many times.

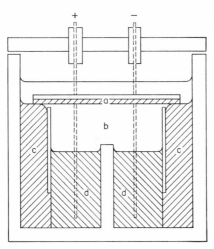

Fig. 56.   Perspex cell for agar electrophoresis. a, microscope slide or lantern plate with agar surface down; b, petroleum ether; c, 0·8% agar in barbitone buffer, pH 8·4, μ 0·05; d, same buffer (reproduced with permission from Blanchaer, 1961a).

B. STARCH-GEL ELECTROPHORESIS

The technique of starch-gel electrophoresis was first described by Smithies (1955; 1959a) and with only minor modifications has since been used extensively in enzyme studies. The partially hydrolysed starch used in making the gels may be prepared according to the technique of Smithies (1955) but the commercially available product specially prepared for starch-gel electrophoresis (Connaught Medical Research Laboratories) is more commonly employed.

(i) *Preparation of Gel.* A suspension of starch in buffer solution is heated in a Pyrex round-bottomed flask. The concentration of starch varies according to the batch, but is usually in the 11–13% range; instructions are given on the label. The effect of varying the starch concentration on the migration of proteins has been investigated and used to estimate the relative sizes of proteins (Smithies, 1963). The size of flask is important as not more than 300 ml of gel should be made in a 500 ml flask. The suspension is heated over a naked flame with continuous agitation until the starch grains are ruptured and a semi-solid opaque mass is formed. Heating is continued until a viscous translucent solution is obtained. A more controlled method of preparing the gel where the starch suspension is heated for a standard time in a boiling water bath with mechanical or manual stirring has been described (Smithies, 1963; Boyde, T. R. C., personal communication). The flask is shaken for a few seconds away from the source of heat and negative pressure applied with a water pump until the contents "boil". The material in the flask is swirled vigorously during the degassing which is continued for 5–10 seconds. The viscous translucent solution is poured into suitable plastic moulds and covered with a glass or Perspex plate. The gels are usually left to set from 3–17 hours at room temperature. Best results are usually obtained with a freshly prepared gel, i.e. less than 5 hours after pouring, although the separations with a gel left overnight only differ in so far as the separated protein zones are a little more diffuse.

Gels for horizontal electrophoresis (Smithies, 1955) are prepared in Perspex trays 25 cm long and 6 mm deep, the width varying from 2–16 cm depending on the number of samples.

For vertical electrophoresis, which is preferable owing to the better resolution and reproducibility (Smithies, 1959a) the gels while setting are covered with a Perspex lid which incorporates a slot former. The gel trays are usually about 30 cm long × 6 mm thick; the width varying as in the horizontal method. The two end plates are best held in place with rubber bands and can be removed easily.

(ii) *Sample Introduction*. The method of sample introduction employed is dependent upon whether horizontal or vertical starch-gel electrophoresis is to be used.

With horizontal electrophoresis, there are two methods for introducing the sample.

(*i*) The first of these is by filter paper insertion. A transverse cut across the whole width of the gel is made with a razor blade and is situated approximately one third of the length from one end, which is usually the cathode. The cut is opened by exerting pressure on the blade. A piece of filter paper (Whatman No. 3), previously cut to fit the cross-section of the gel, is held by forceps and immersed in the sample. The moist filter paper is inserted in the cut so that it adheres to the undisplaced surface of the gel. The slot is closed by applying slight pressure on the gel so as to return it to its original position, care being taken to avoid air-bubbles. Multiple samples can be applied by using shorter pieces of filter paper.

(*ii*) The second method of sample application is known as starch insertion. A suspension of starch grains in the sample under investigation is drawn up into a Pasteur pipette and allowed to settle in a vertical position. A slight excess of the suspension is transferred to a slot previously made in the gel by cutting out a small block with two razor blades mounted 2–3 mm apart. The slot is situated, once again, approximately one third of the length along the gel. The excess of supernatant sample is removed by blotting with filter paper, more material added and the process repeated until a uniformly packed block of starch grains is obtained. Here again, multiple samples can be applied by making smaller slots. For this purpose a slot former can be employed.

After applying the sample or samples, molten petroleum jelly at approximately 45 °C is poured over the gel leaving 2–3 cm uncovered at each end for the electrical connections. The gel is assembled as in Fig. 57.

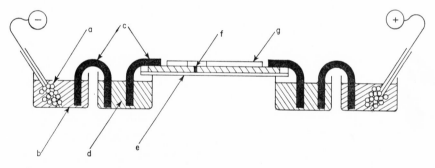

Fig. 57. General layout of apparatus for horizontal electrophoresis. a, Ag/AgCl electrode; b, concentrated NaCl solution (10%) in electrode compartment; c, filter paper bridges soaked in bridge solution (e.g. 0·3 M borate, pH 8·6); d, compartment containing bridge solution; e, starch gel contained in Perspex mould; f, position of sample insertion; g, petroleum jelly seal to prevent loss of water during electrophoresis (reproduced with permission from Smithies, 1955).

With vertical electrophoresis, the sample slots are pre-formed in the gel during the setting stage. The slot former may be made of one-sixteenth inch or one-thirtysecond inch thick Perspex or of portions of glass microscope slides 1·5 mm thick cemented into a Perspex shoulder. Glass inserts may also be held in position with "Parafilm" which makes the number of samples examined in the one gel easily variable. The samples are added directly into the slots with a Pasteur pipette and

petroleum jelly at approximately 45°C is poured over the slots to seal
in the samples. The whole surface of the gel is then covered with molten
vaseline. The end plates of the gel tray are removed and the gel
assembled in a vertical position with the base supported on a thick wad
of filter paper in the bottom buffer compartment (Fig. 58).

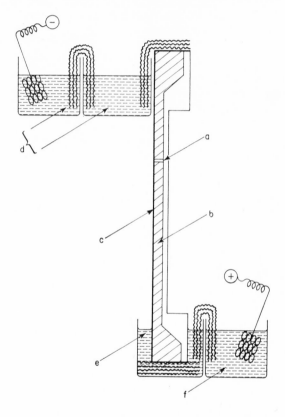

Fɪɢ. 58.   Arrangement for vertical starch-gel electrophoresis. a, position of sample slots;
b, gel; c, petroleum jelly seal; d, tray containing bridge solution (e.g. 0·3 ᴍ borate, pH 8·6);
e, tray containing bridge solution with a number of thicknesses of filter paper at the bottom
on which end of gel rests; f, tray containing concentrated NaCl solution (10%) (reproduced
with permission from Smithies, 1959a).

For enzyme studies the electrophoresis is usually carried out at 4°C.
The gels are pre-cooled for at least 1½ hours before introduction of the
sample.

For horizontal electrophoresis a voltage gradient of 6–8 V/cm is
applied for about 6 hours (Smithies, 1955).

For vertical electrophoresis a voltage gradient of 4–5 V/cm may be applied for 16–18 hours (Smithies, 1959a) but as enzyme activity is usually reduced during a long electrophoresis run it is often preferable to use a voltage gradient of 10 V/cm for $1\frac{1}{2}$–3 hours (Latner and Skillen, 1961). In practice the voltage gradient used is often determined by the type of power supply available and the length of time adjusted by trial and error until good separation is obtained.

(iii) *Buffer Solutions.* The original starch-gel electrophoresis technique was developed for serum proteins using 0·025 M borate buffers at pH 8·6 (Smithies, 1955; 1959b). Similar borate buffers have been used in studies of esterases (Hunter and Markert, 1957) and lactate dehydrogenase (Allen, J. M., 1961). Tris–HCl buffers were recommended for dehydrogenases (Tsao, 1960; Latner and Skillen, 1961) and alkaline phosphatases (Hodson et al., 1962). Barbitone buffers have been used for lactate dehydrogenase (Wróblewski and Gregory, 1961) and isocitrate dehydrogenase (Baron and Bell, 1962). Other buffers used for separations of lactate dehydrogenases are EDTA/borate/Tris (Boyer et al., 1963; Nance et al., 1963) and phosphate–citrate (Philip and Vesell, 1962; Fine and Costello, 1963). The ionic strengths and pH values of the various buffer systems used are given in Table 13. Normally the bridge buffers are the same composition as the gel buffers but about ten times more concentrated, although discontinuous systems in which the gel and bridge buffers differ radically in composition are sometimes used. The discontinuous buffer system of Poulik (1957) has been used for example in studies on alkaline phosphatase (Moss and King, 1962) and esterases (Paul and Fottrell, 1961). The importance of buffer conditions in the starch-gel electrophoresis separation of muscle proteins (Neelin, 1963) and lactate dehydrogenases (Ressler et al., 1963a; 1963b) has been discussed.

C. ACRYLAMIDE GELS

The use of synthetic acrylamide gels as an electrophoresis medium has been introduced during the last 6 years (Raymond and Weintraub, 1959; Hermans et al., 1960; Raymond and Wang, 1960; Ornstein and Davis, 1961; Raymond, 1964). Acrylamide is completely soluble in buffer solutions and polymerizes to form a suitable supporting medium for gel electrophoresis. In contrast to starch gels, the polyacrylamide gels are thermostable, transparent, strong and relatively chemically inert; they can be prepared with varying pore sizes (Ornstein and Davis, 1961). Using acrylamide gels, very high resolution of proteins may be obtained in relatively short runs.

H*

TABLE 13

Buffer solutions for starch-gel electrophoresis of enzymes

Enzyme	Gel buffer	Bridge buffer	Gel pH	Ref
Oxidoreductases				
Lactate dehydrogenase	0·025 M H_3BO_3/0·01 M NaOH	0·3 M H_3BO_3/0·06 M NaOH	8·6	Markert and Moller, 1959
	0·03 M Tris-HCl	0·05 M Tris-HCl	8·4	Tsao, 1960
	0·05 M Tris-HCl	0·3 M Tris-HCl	8·6	Latner and Skillen, 1961a
	0·01 M Na_2HPO_4/citrate	0·2 M Na_2HPO_4/citrate	7·0	Philip and Vesell, 1962
	0·01 M barbitone/0·05 M Na barbitone	0·02 M barbitone/0·1 M Na barbitone	8·6	Wróblewski and Gregory, 1961
	0·09 M Tris/0·05 M H_3BO_3/0·002 M EDTA	0·11 M Tris/0·06 M H_3BO_3/0·0024 M EDTA	8·5	Nance et al., 1963
Malate dehydrogenase	0·0086 M Na_2HPO_4/citrate	0·032 M Na_2HPO_4/citrate	7·0	Thorne et al., 1963
Isocitrate dehydrogenase	0·03–0·04 M Na_2HPO_4/citrate	0·3 M Na_2HPO_4/citrate	7·0	Henderson, 1965
	0·005 M citrate	0·2 M citrate	6·2	Campbell and Moss, 1962
Glucose-6-phosphate dehydrogenase	0·076 M Tris/0·005 M citrate	0·3 M H_3BO_3/0·05 M NaOH	8·6	Boyer et al., 1962
	0·05 M Tris-HCl/0·0027 M EDTA	0·05 M Tris-HCl/0·05 M NaCl/0·0027 M EDTA	8·8	Kirkman and Hendrickson, 1963
Phosphogluconate dehydrogenase	0·01 M phosphate	0·1 M phosphate	7·0	Fildes and Parr, 1963
Alcohol dehydrogenase	0·025 M H_3BO_3/0·01 M NaOH	0·3 M H_3BO_3/0·06 M NaOH	8·5	Schwartz, 1965
Galactose dehydrogenase	0·0053 M phosphate	0·053 phosphate	6·7	Shaw, 1966
Catalase	0·076 M Tris/0·005 M citrate	0·3 M H_3BO_3/0·06 M NaOH	8·6	Paul and Fottrell, 1961
	0·025 M H_3BO_3/0·01 M NaOH	0·3 M H_3BO_3/0·06 M NaOH	8·6	Markert and Moller, 1959

	Gel buffer	Electrode buffer	pH	Reference
Transferases				
Aspartate aminotransferase	0·025 M H_3BO_3/0·01 M NaOH	0·3 M H_3BO_3/0·06 M NaOH	8·6	Boyde and Latner, 1962
	0.76 M Tris/0·005 M citrate	0·3 M H_3BO_3/0·06 M NaOH	8·6	Schwartz et al., 1963
Phosphoglucomutase	0·01 M Tris/0·01 M maleic acid/0·001 M EDTA/0·001 M $MgCl_2$/NaOH	0·1 M Tris/0·01 M maleic acid/0·01 M EDTA/0·01 M $MgCl_2$/NaOH	7·4	Spencer et al., 1964
Adenylate kinase	0·005 M histidine/NaOH	0·41 M citric acid/NaOH	7·0	Fildes and Harris, 1966
Hexokinase	0·02 M Na barbital/0·027 M EDTA/0·005 M mercapto-ethanol	As gel	8·4	Katzen and Schimke, 1965
Hydrolases				
Alkaline phosphatase	0·076 M Tris/0·005 M citrate	0·3 M H_3BO_3/0·05 M NaOH	8·6	Boyer, 1961
	0·05 M Tris-HCl	0·3 M Tris-HCl	8·6	Hodson et al., 1962
Acid phosphatase	0·025 M H_3BO_3/0·01 M NaOH	0·3 M H_3BO_3/0·06 M NaOH	8·6	Estborn, 1959
	0·005 M Tris/citrate	0·2 M Tris/citrate	6·2	Sur et al., 1962
Esterases	0·025 M H_3BO_3/0·01 M NaOH	0·3 M H_3BO_3/0·06 M NaOH	8·6	Hunter and Markert, 1957
	0·076 M Tris/0·005 M citrate	0·3 M H_3BO_3/0·06 M NaOH	8·6	Paul and Fottrell, 1961
Leucine aminopeptidase	0·025 M H_3BO_3/0·01 M NaOH	0·3 M H_3BO_3/0·06 M NaOH	8·6	Dubbs et al., 1960
Lyases				
Aldolase	0·002 M EDTA/0·05 M H_3BO_3/0·09 M Tris	0·002 M EDTA/0·06 M H_3BO_3/0·11 M Tris	8·6	Anstall et al., 1966

(i) *Preparation of Gel.* There are two different methods of preparation of the gels, one which is used for the simple block gel (Raymond and Wang, 1960) and a second which involves disc electrophoresis in columns of gel consisting of three sections (a) a large pore anti-convection gel containing the protein sample, (b) a large pore spacer gel in which electrophoretic concentration takes place and (c) a small pore gel in which the electrophoretic separation takes place (Ornstein and Davis, 1961; Reisfeld *et al.*, 1962).

For the simple block type of gel (Raymond and Wang, 1960) 5 g of acrylamide monomer (Cyanogum-41, American Cyanamid Co.) is dissolved in 100 ml of the desired buffer. 0·1 ml of a freshly prepared 10% solution of dimethyl aminopropionitrile in the same buffer is now added followed by 1 ml of freshly prepared 10% ammonium persulphate in aqueous solution. The mixture is poured into a suitable mould and allowed to set for 3 hours or longer. If the monomer-catalyst solution is exposed to the air, the top layer to a depth of 1 cm does not polymerize and the gel mould must be designed so that the whole gel surface is covered. The type of mould used for vertical starch-gel electrophoresis is most suitable and the subsequent set-up for electrophoresis is also very similar to the vertical starch gel.

For the technique of disc electrophoresis (Ornstein and Davis, 1961; Reisfeld *et al.*, 1962) five stock solutions are employed. They are:

(a) 100 ml buffer containing 4·0 ml tetramethylethylenediamine (TEMED).

(b) 100 ml buffer containing 0·46 ml TEMED.

(c) 60 g acrylamide and 0·4 g methylenebisacrylamide (BIS) dissolved in 100 ml distilled water.

(d) 10 g acrylamide and 2·5 g BIS dissolved in 100 ml distilled water.

(e) 4 mg riboflavin dissolved in 100 ml distilled water.

The buffers employed have been Tris–glycine pH 8·9 (Ornstein and Davis, 1961) and β-alanine acetate pH 4·3 (Reisfeld *et al.*, 1962), although buffers for starch gels may be used.

The small pore gel is prepared by mixing one part of (a), two parts of (c) and one part of distilled water with four parts of a freshly prepared aqueous solution of ammonium persulphate (0·28 g/100 ml).

Glass tubes 7 × 0·5 cm internal diameter are tightly closed at one end with flat-topped rubber bungs. The tubes are placed upright on the bungs and filled to 0·5 in from the top with the small pore solution. This solution is carefully overlaid with a 0·25 in column of water, care being taken to avoid distortion at the interface. The tubes are left at room temperature for 30 minutes for the gel to form.

The large pore gel is prepared by mixing one part of (b), two parts of (d), one part of (e) and four parts of distilled water. The water layer is removed from the tubes and the large pore gel solution added to about 0·125 in from the top. A water layer is again overlaid and the tubes placed within 6 inches of a 15 W fluorescent lamp for about 15 minutes to photopolymerize. After the large pore gel has polymerized, the sample is applied to the top of the tube using a mixture of 0·15 ml large pore gel solution and 5–10 μl sample (50–200 μg protein) and photopolymerized for 20 minutes (no water layer is overlaid). It has been shown that if a slurry of the protein sample is made with Sephadex G50 (Pharmacia, Sweden) the large pore gel is unnecessary (Racusen and Calvanico, 1964). After addition of the sample, the tubes are carefully filled up with electrode buffer and assembled as in Fig. 59. A voltage gradient of 20 V/cm is applied for about 30 minutes at room temperature.

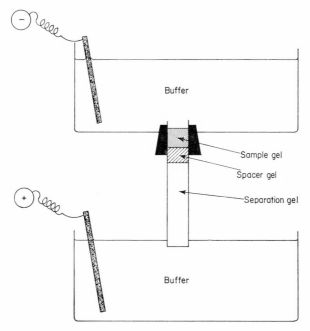

FIG. 59. Disc electrophoresis assembly. The tube containing the acrylamide gel is held in the upper buffer compartment with a rubber bung.

Comparative studies of starch-gel and acrylamide-gel electrophoresis of serum proteins have shown that the acrylamide system of electrophoresis would appear to be more sensitive and that most of the increased sensitivity is due to the diminished depth of the sample zone, which is produced by electrophoretic concentration (Ornstein and Davis, 1961). Some workers (Allen and Hynick, 1963) have shown that several enzymes which did not separate easily in starch gel, e.g. "insoluble" alkaline phosphatases, nucleoside diphosphatases, thiamine pyrophosphatases, rat brain cholinesterases and NAD and NADP diaphorases, were easily separated in acrylamide gel, although acid phosphatases from various sources could be separated only after starch-gel electrophoresis.

D. IMMUNOELECTROPHORESIS

This combines the Ouchterlony double diffusion technique with separation by electrophoresis (Grabar and Williams, 1953; Grabar, 1958). It is most commonly carried out in agar alone but various workers have combined the superior separation of starch-gel electrophoresis with double diffusion in agar (Poulik, 1959; Butler and Flynn, 1961; Latner and Skillen, 1961a; Schleuren and Goll, 1961; Press and Porter, 1962).

After the electrophoretic separation of the antigens, a trough is cut in the agar gel parallel to the axis of electrophoretic separation. This trough is filled with antiserum and the agar plate is then incubated in a "damp-box" at room temperature until the precipitin lines are fully developed. The agar plates are then washed with physiological saline or buffer solution to remove excess antigen and antibody, leaving the insoluble precipitin pattern which can be examined for enzyme activity by the usual techniques.

E. CELLULOSE ACETATE ELECTROPHORESIS

Cellulose acetate was introduced and developed as a medium for electrophoresis by Kohn (1957a; 1957b; 1960). Better separations can be obtained within a very much shorter time than with conventional paper electrophoresis. Very small quantities of protein may be employed, adsorption and subsequent "trailing" is minimal and the membranes are homogeneous and chemically relatively pure. Electrophoresis is carried out in a horizontal tank.

The cellulose acetate strips are cut to size and the strips floated on the surface of some electrophoresis buffer. This allows the buffer to soak up evenly into the membrane strip. When white opaque areas no longer show, the strips are removed from the buffer and lightly blotted with filter paper. Using forceps, the strips are transferred to the electrophoresis tank and assembled in position. The strip-holders are used to hold the cellulose acetate strips reasonably taut so that the strips do not sag in the centre. If necessary, a support may be used for long strips.

The best support is provided by a series of plastic pins at intervals along the central partition of the electrophoresis tank. The sample is applied with a capillary or micro-pipette by moving the capillary along the edge of a ruler. The tip of the capillary is rounded in a flame to avoid scratching and the flow controlled with finger tip pressure on the top of the capillary. A 5 mm margin is left at either side of the strip.

Sample volumes up to 10 μl may be applied in this manner and 5 μl is sufficient for separations of human serum proteins. The separation pattern is controlled by the position of the starting line. For most purposes the starting line is usually one third the length of the strip from the cathode end.

Either a constant voltage or constant current is applied with a maximum current of 0·5 mA/cm width of strip. The buffer recommended for serum proteins is 0·07 M barbitone buffer pH 8·6. For the more usual type of cellulose acetate electrophoresis, the strips are 10–12 cm long and the voltage is applied for about 2 hours.

After electrophoresis the acetate strips are removed from the tank and may be either cut up into strips and eluted (Wieland *et al.*, 1959a) with subsequent enzyme analysis of the fractions, or visual staining techniques may be applied to the intact strip. For this purpose, a moist chamber is prepared and a clean glass slide the same size as the electrophoresis strip is placed in the bottom. A small quantity of staining medium is pipetted on to the slide (0·25 ml) and the electrophoresis strip floated on this in the same way as the strip was wetted with buffer before electrophoresis. When the strip is fully impregnated with stain, it is incubated at 37°C for $\frac{1}{2}$–1 hour.

Isoenzymes of lactate dehydrogenase (Barnett, 1962; 1964), creatine phosphokinase (Rosalki, 1965) and leucine aminopeptidase (Meade and Rosalki, 1964) have been rendered visible in this manner.

F. COLUMN ELECTROPHORESIS

Electrophoresis in columns of cellulose powders has been employed in investigations of plasma esterases (Augustinsson, 1958; 1961) and serum aspartate aminotransferases (Augustinsson and Erne, 1961).

The electrophoresis is normally carried out in a column 40 cm long and 3 cm in diameter. A commercially available apparatus has been used in most of the investigations of serum enzymes (LKB Column Electrophoresis Apparatus 3340C). The column is provided with a cooling jacket and ground glass joints at both ends for connection to the anode and cathode electrode vessels.

The electrophoresis runs are carried out in barbitone buffer (pH 8·4, $\mu = 0·1$). The temperature is kept constant within the range 5–11°C.

A voltage of 260 V with a current of 60 mA is applied for 30 hours. For each electrophoresis 5 ml of buffered plasma or serum previously dialyzed against the electrophoresis buffer is layered carefully on to the top of the column. The sample zone is allowed to move 2–3 cm below the upper surface of the cellulose column before applying the voltage gradient.

After completion of the electrophoresis, the protein fractions are displaced from the column with buffer. The flow rate is approximately 10 ml/h, with fraction volumes of 3 ml.

G. CHROMATOGRAPHY ON DEAE CELLULOSE

The use of column chromatography on DEAE cellulose as a means of protein fractionation was introduced by Sober and Peterson (1954) and applied to serum proteins 2 years later (Sober *et al.*, 1956).

Following this success, similar techniques have been applied to a number of enzymes, starting with the separation of serum and tissue lactate dehydrogenase isoenzymes (Hess and Walter, 1960; 1961).

(i) *Preparation of the Cellulose.* The DEAE cellulose is suspended in 0·008 M sodium phosphate buffer pH 7·0. After stirring for 15 minutes, the cellulose is washed two or three times by decantation with the same buffer so that the fines are removed.

(ii) *Column Chromatography.* The prepared cellulose is suspended in the buffer and made into a suitable column.

Before being applied to the column, the serum or tissue extract under investigation is dialyzed for 4 hours at 4°C against the 0·008 M phosphate buffer.

A 5–10 ml sample may be applied to a column 20 cm long and 8 mm in diameter. The following series of buffers have been used in separations of lactate dehydrogenase on a column of these dimensions with a flow rate of about 10 ml/h (Hess and Walter, 1960). (a) 45 ml 0·008 M sodium phosphate buffer pH 7·0, (b) 45 ml 0·010 M sodium phosphate buffer pH 6·0, (c) 45 ml 0·020 M sodium phosphate buffer pH 6·0, (d) 70 ml 0·05 M sodium phosphate buffer pH 6·0, (e) 70 ml 0·05 M sodium phosphate buffer pH 6·0/0·02 M NaCl, (f) 100 ml 0·05 M sodium phosphate buffer pH 6·0/0·05 M NaCl, (g) 150 ml 0·10 M sodium phosphate buffer pH 6·0/0·10 M NaCl, (h) 100 ml 0·10 M sodium phosphate buffer pH 6·0/0·10 M NaCl, (i) 100 ml 0·20 M sodium phosphate buffer pH 6·0/0·20 M NaCl, (j) 100 ml 0·40 M sodium phosphate buffer pH 6·0/0·40 M NaCl.

These solutions are applicable to separations of many different enzymes. In some cases it is more helpful to use a continuous NaCl gradient elution up to a final concentration of 0·3 M NaCl.

(iii) *Batch technique.* A simple batch technique for the quick differentiation between the "heart" and "liver" isoenzymes LDH-1 and LDH-5 was also described by Hess and Walter (1961).

Two millilitres of serum which has been dialyzed for 2 hours against 0·02 M phosphate buffer pH 6·0 are mixed with 2 ml of a 10% suspension of DEAE cellulose in the same buffer. After stirring gently for 5 minutes the suspension is centrifuged and the lactate dehydrogenase activity of the supernatant is determined. The difference in the activity of the supernatant and the original serum represents the activity of those isoenzymes which are most prominent in tissue extracts of heart muscle as they are adsorbed on to the DEAE cellulose.

(iv) *DEAE Sephadex.* Following the results of Hess and Walter (1960; 1961) a modification has been introduced which employs DEAE Sephadex in place of DEAE cellulose (Richterich *et al.*, 1963). It has been suggested that the DEAE Sephadex is a more consistent product than DEAE cellulose and gives more reproducible results.*

The resin is prepared by soaking in distilled water overnight to swell the dextran grains. The Sephadex gel is then washed with water three times by decantation to remove fines. The Sephadex is stirred for 5 minutes with 0·5 M HCl and then washed with distilled water. After resuspending in 0·5 M NaOH and stirring for a further 5 minutes the excess alkali is washed off with excess distilled water. The pH of the gel suspension is adjusted to pH 7·5 with 0·1 M HCl and the gel then suspended in 0·05 M triethanolamine–HCl–0·005 M EDTA buffer pH 7·5. The gel is washed with two changes of the same buffer and is then ready for use.

DEAE Sephadex may be used in a column as for the DEAE cellulose or it may be used in the adsorption–elution technique of Richterich and co-workers (1963).

In this technique 0·1 ml of serum or tissue extract is mixed with 0·3 ml DEAE Sephadex suspension (one part gel to two parts 0·05 M buffer) and stirred gently for 1 minute. The suspension is centrifuged and 0·2 ml of the supernatant is removed and assayed—this contains LDH-5. 0·2 ml of 0·05 M buffer containing 50 mmoles/l NaCl is added to the remaining suspension and the centrifugation repeated. LDH-4 is removed in the supernatant.

* DEAE Sephadex has been used in the preparation of crystalline LDH isoenzymes from human tissues (Wachsmuth and Pfleiderer, 1963).

The whole process is repeated using 0·05 M buffer containing 100 mmole/l and 300 mmole/l NaCl removing firstly the LDH-3 and then a mixture of the LDH-1 and LDH-2 isoenzymes.

H. GEL FILTRATION

Gel filtration is a means of separating substances on chromatographic columns by virtue of their molecular weight and shape (Flodin, 1962). The molecular sieve is usually a cross-linked dextran (Sephadex AB Pharmacia, Uppsala, Sweden), but a polyacrylamide gel has recently become available in a similar form to Sephadex (Bio-gel P, Bio-Rad Laboratories, California). Agar granules have also been used in a gel filtration technique (Polson, 1961; Andrews, 1962).

As Sephadex has been the most widely used medium, this methodology will be discussed in detail. The sizes of Sephadex of interest in separation of enzymes are G75, which excludes molecules of approximately 50,000, G100, G150 and G200, which exclude molecules with molecular weights approximately 100,000, 150,000 and 200,000, respectively.

The dextran is allowed to swell in a large volume of water for 24–72 hours before being packed into a column with a diameter-to-height ratio of at least 1:10. The easiest way of packing is to attach a large funnel to the top of the actual column. The column is filled with buffer and the funnel with the Sephadex suspension. The buffer is allowed to flow slowly out of the column as the Sephadex grains sink down it. In some instances, the Sephadex contains some fines, but these may be removed by repeated suspension and decantation before adding the suspension to the column. When the suspension has settled to the required level, the column is washed through with at least two volumes of the buffer to be used in the separation. A filter paper disc of the same diameter as the column is allowed to drop on to the top of the gel. The sample is applied in as small a volume as possible by carefully layering on to the top of the column. This may be done underneath the buffer solution by adding sucrose to the sample and using a pipette with the tip bent at right angles. As soon as the sample application is completed, buffer solution is allowed to flow down the column and the enzyme fractions are eluted.

I. OTHER TECHNIQUES

(i) *Heat Stability*. Wróblewski and Gregory (1961) have developed a simple test for clinical purposes which utilizes the differential heat stabilities of the lactate dehydrogenase isoenzymes (Plagemann *et al.*, 1961). Two millilitres serum and 0·2 ml NADH$_2$ solution (2·5 mg/ml) are mixed and left for 20 minutes. A 0·5 ml portion of the mixture is now pipetted into three tubes, one of which is kept at room temperature and the other two heated for 30 minutes at 57°C and 65°C, respectively. The tubes are then cooled to room temperature and all three tubes assayed in the normal manner. The activity of the unheated sample represents the total enzyme activity; the difference in activities between the unheated sample and that heated at 57°C represents the heat-labile enzyme, which is elevated in liver disease, and the activity remaining after heating at 65°C represents the heat stable enzyme, which is elevated after myocardial infarction. The "heat stability index" proposed by Latner and Skillen (1963) represents the ratio of activity of serum heated at 60°C for 1 hour to that of unheated serum and is apparently a good indication of myocardial infarction.

(ii) *Effect of Organic Solvents.* The sensitivity of the slower moving serum lactate dehydrogenase isoenzymes to acetone (Latner and Turner, 1963) or chloroform (Warburton *et al.*, 1963) has also been suggested as an aid to the estimation of the "heart" lactate dehydrogenase isoenzymes in serum. The former test involves mixing 0·5 ml acetone, 0·5 ml water and 0·5 ml serum and leaving for 10 minutes at room temperature. After centrifugation 0·3 ml of supernatant is evaporated to dryness in a vacuum desiccator. The residue is reconstituted in 0·5 ml 0·1 M phosphate buffer ph 7·0 and assayed in the normal manner. A supernatant level of 35 m.I.U./ml is suggestive of myocardial infarction, although false-positive results are given by certain cases of cancer and untreated pernicious anaemia.

Precipitation of serum alkaline phosphatase with alcohol has been suggested as a means of differentiating between bone disease and liver disease (Peacock *et al.*, 1963). One millilitre of a tenfold dilution of serum is mixed at 0°C with 4 ml of 25% ethanol in 0·025 M acetate buffer pH 5·0 and after leaving for 15 minutes at 0°C the precipitate is centrifuged off in the cold. The precipitate is dissolved in 1 ml of buffered substrate and assayed in the normal manner. Only very small amounts of bone alkaline phosphatase are precipitated.

(iii) *Pyruvate and lactate inhibition, and coenzyme analogues as a means of differentiating LDH isoenzymes.* Plagemann *et al.* (1960b) have described different pH optima for the lactate dehydrogenase activity of various human and rabbit tissues. Estimation of LDH activity using 1·2 mM pyruvate and 0·15 mM pyruvate can be used as an indication of the relative proportions of the isoenzymes in a particular enzyme preparation. Tissues such as liver and skeletal muscle with a preponderance of the electrophoretically slower moving isoenzymes have ratios of the activity at these two substrate concentrations of 1·5–2·3, whereas tissues such as heart, kidney and erythrocytes, which contain more of the faster moving isoenzymes, have ratios of 0·38–0·44 (Plagemann *et al.*, 1960b).

The ratio of LDH activity, using 0·33 mM and 10 mM pyruvate as substrates, has also been used to indicate whether a particular preparation contains more of the M or H sub-units (Cahn *et al.*, 1962; Wilson *et al.*, 1963; Salthe, 1965). The ratio of the activity at 0·33 mM and 10 mM pyruvate is usually above 2·0 for the enzyme from tissues containing a preponderance of H sub-units and below 2·0 for the enzyme from tissues with a preponderance of M sub-units.

Stambaugh and Post (1966c) have described a spectrophotometric method for the assay of the amounts of "H" and "M" sub-units of lactate dehydrogenase in a particular preparation of the enzyme from either human or rabbit sources. The method depends on a differential assay at high (250 mM) and low (15 mM) lactate concentrations. For the human enzyme the M_4 isoenzyme has maximal activity at approximately 250 mM lactate, whereas the H_4 isoenzyme is maximally active at 10 mM lactate and is inhibited by approximately 50% at 250 mM lactate. If the change in OD at 340 nm with 15 mM lactate as substrate is ΔA_{low} and the change with 250 mM lactate is ΔA_{high} then $\Delta A_{low} = 0·965 \Delta A_H + 0·45 \Delta A_M$ where ΔA_H and ΔA_M represent the change in OD for the two sub-units at their optimal substrate concentration. Also $\Delta A_{high} = 0·49 \Delta A_H + \Delta A_M$. This means that $\Delta A_M = 1·3 \Delta A_{high} - 0·659 \Delta A_{low}$ and $\Delta A_H = 1·34 \Delta A_{high} - 0·605 \Delta A_{low}$.

From these formulae the amounts of M and H sub-units can easily be calculated.

When a variety of coenzyme analogues are used in the assay of lactate dehydrogenase activity it is found that the activity varies according to which analogue is used (Kaplan *et al.*, 1960). The enzymes from different sources can sometimes be identified by measuring the ratio of the LDH activity of a particular preparation with different coenzyme analogues (Kaplan and Ciotti, 1961).

(iv) *The use of substrate analogues in the estimation of lactate dehydrogenase isoenzymes.* Rosalki and Wilkinson (1960) have shown that if 3·3 mM α-oxobutyrate is used in place of 0·33 mM pyruvate for the spectrophotometric assay of LDH activity at pH 7·4 the isoenzymes exhibit different affinities for the two substrates, LDH-1 being much more active with oxobutyrate than pyruvate, whereas LDH-5 is much more active with the latter substrate. Assay of an enzyme preparation with these concentrations of the substrates allows the relative proportions of the LDH isoenzymes present to be determined. The ratio of the serum enzyme activities with oxobutyrate and pyruvate is a useful index which has been employed as a diagnostic aid (Elliot *et al.*, 1962; Elliot and Wilkinson, 1963). The HBD/LDH ratio for normal serum is within the range 0·63–0·81. A ratio greater than 0·83 is usually found after myocardial infarction and a ratio lower than 0·63 is consistent with an hepatic disorder.

(v) *Urea and oxalate inactivation of lactate dehydrogenase.* LDH-4 and LDH-5 are inhibited by much lower concentrations of urea than LDH-1 and LDH-2 (Richterich and Burger, 1963; Brody, 1964; Withycombe *et al.*, 1965). A simple method for demonstration in serum of the LDH isoenzymes from heart muscle has been developed by Hardy (1965) using a spectrophotometric assay system containing 2·6 M urea which inhibits the slower moving isoenzymes.

Emerson and Wilkinson (1965) have reported that urea and oxalate inhibition are useful means of differentiating between serum lactate dehydrogenase isoenzyme patterns. Incorporation of 2 M urea or 0·2 mM oxalate into a standard spectrophotometric assay system for lactate dehydrogenase reduced normal serum activity by about 50%. Similar concentrations of these inhibitors in an α-hydroxy-butyrate dehydrogenase system reduces normal serum enzyme activity by 80% with 2 M urea and 5% with 0·2 mM oxalate. Samples containing more of LDH-1 and LDH-2 will have urea inhibition less than 45% and oxalate inhibition greater than 70%, whereas those samples with more LDH-4 and LDH-5 will have urea inhibition greater than 60% and oxalate inhibition less than 50%.

2. METHODS OF DEMONSTRATING ENZYME ACTIVITY

Enzyme activity is demonstrated in a variety of ways. The material may be extracted from appropriate segments and the assay of the extracts carried out by classical techniques. For reactions involving proton transfer from $NADH_2$ or $NADPH_2$ to NAD or NADP, bands of activity can be demonstrated by ultraviolet light illumination. With other enzymes, for example alkaline phosphatase, it is possible to use a non-absorbing substrate, which as a result of enzyme activity liberates a substance which absorbs in the ultraviolet region of the spectrum. There are various techniques for rendering visible the region or regions where enzyme activity exists. This can be done by specific staining techniques or the utilization of immunological procedures. In either case, photography gives a permanent record. Scanning by transmission or reflectance can also be used for recording after ultraviolet light illumination or direct visual staining.

Quantitation of gel electrophoresis, even with simple protein stains, is difficult to achieve as dye binding is not uniform throughout the range of protein fractions. Using histochemical stains for enzymes, the problems include diffusion of the substrate into the gel matrix. If a precipitate is deposited on the gel surface by enzyme action, this precipitate may retard diffusion of more substrate further into the gel and if there is a large amount of enzyme activity located in any particular zone, the resultant rapid rate of reaction will most probably prevent deposition of the histochemical precipitate within the surface of the gel as the products of the enzyme reaction will simply be diffused into the general substrate solution. It has been noted that zones of high enzyme activity when visualized directly on starch gels often appear with an unstained central portion of the zone. Other difficulties arise in so far as the isoenzymes may have different substrate affinities. This means that, although two isoenzymes may have exactly the same activity when estimated under optimal conditions, they are unlikely to show the same activity when visualized by a direct staining technique. The width of the zone of enzyme activity in the gel may have marked effects on the intensity of histochemical staining. Some isoenzymes always travel as a broader band than others and the amount of staining will depend on the width of the zone of enzyme activity.

Enzyme resolution during starch-gel electrophoresis has been examined by White and Kushnir (1966) with special reference to the α-glucosidases in honey. The results indicate that for the purposes of research it is often better to estimate enzyme activity in serial segments of the gel rather than visualize the enzyme with a chromogenic substance. With the example of α-glucosidase only one zone of the enzyme was found with the visual staining technique, whereas up to ten distinct fractions could be detected after analysis of serial segments.

It is possible to slice starch gels after electrophoresis by means of a cutting block and a fine wire. As many as five slices may be obtained from a standard 6 mm thick gel. The technique employed is essentially that of Smithies (1955) but a fine wire, in the manner of a cheese cutter, is used in preference to a knife. Each slice may be examined for a different enzyme activity and this has decided advantages. In any case, it is advisable to have a relatively thin slice for any staining technique. Non-uniformity of migration throughout the thickness of the gel is a particular problem if serial slices of the gel are stained for different enzymes or proteins whose migration rates are to be compared. Hori (1966) has shown that the methods used for moulding starch gels are important as some distortion of the protein zones is produced if the gels are moulded in the usual manner. If, however, "Saran Wrap" is placed over the top and bottom of the mould during the "setting" period, the gel formation is more homogenous and the protein zones produced during electrophoresis are less likely to be distorted.

A. OXIDOREDUCTASES

(i) *Lactate Dehydrogenase.* There are a number of different methods for demonstrating activity.

Extraction Methods. After electrophoresis the gel is cut into 0·5 cm segments. The enzymes may be extracted from the gel segments by repeated freezing and thawing or homogenization in buffer as for proteins (Smithies, 1955). Using this procedure cabbage enzymes were the first to be studied (Smithies and Dixon, 1957). Elution from agar gel by similar simple extraction methods has been used for examination of brain lactate dehydrogenases (Bonavita and Guanari, 1962; 1963a; 1963b). Wróblewski and Gregory (1961) in their investigations of human and rabbit lactate dehydrogenases in starch gel have subjected the gel segments to freezing and thawing, which unfortunately destroys some enzyme activity, and then macerated the resultant sponge in an equal volume of a solution containing 1 mg/ml α-amylase and 0·067 M phosphate buffer pH 7·0 which was 10% saturated with ammonium sulphate. After centrifugation, the supernatants have been assayed for enzyme activity.

Demonstration by Ultraviolet Light. Dehydrogenases and other enzymes, such as aminotransferases, which can be coupled to dehydrogenase reactions may be demonstrated after gel electrophoresis using the absorbence of $NADH_2$ at 340 nm.

This method has been successfully developed by Wieme (1959a) in his technique of enzymoelectrophoresis. After agar electrophoresis on a microscope slide a second agar layer containing the substrate is laid on top of the electrophoresis plate. For detection of lactate dehydrogenase the substrate plate is prepared as follows (Wieme and Van Maercke, 1961):

Two millilitres of a 2% solution of agar in barbitone buffer (pH 8·4) is heated with 0·5 ml distilled water until the agar dissolves. The solution is cooled to 50°C and 0·1 ml 0·1 M pyruvate and 1 ml $NADH_2$ solution (2 mg/ml) added. The mixture is then poured on to a microscope slide and allowed to set. The twinned plates can be transferred to a spectrophotometer and scanned at 340 nm or photographed under ultraviolet light after about 30 minutes. By scanning at set time intervals enzyme kinetics may be studied. The substrate may be applied directly to the agar electrophoresis plate either as a thin film or on a moistened filter paper strip (Blanchaer, 1961a). The ultraviolet light technique has also been used with starch gel (Blanchaer, 1961b) but as this gel is not transparent, photographic recording by reflection must be employed. A similar ultraviolet fluorescence technique for rendering lactate dehydrogenase isoenzymes visible after high voltage cellulose acetate electrophoresis (40 V/cm) has been developed (Pfleiderer and Wachsmuth, 1961).

Visual Staining Technique. Although the use of methylene blue and diaphorase as electron transfer agents in techniques for demonstrating lactate dehydrogenase after starch-gel electrophoresis yielded reasonable results (Markert and Møller, 1959; Tsao, 1960) the substrate had to be incorporated in a starch- or agar-gel medium and incubation took up to 24 hours before the isoenzyme pattern was developed. The introduction of phenazine methosulphate (Dewey and Conklin, 1960; Latner and Skillen, 1961; Van der Helm, 1961) led to much superior results in shorter times and the agar- or starch-gel base need not be used. This compound and the newer tetrazolium salts MTT (3-(4,5-dimethyl-thiazolyl-2)-2,5 diphenyl tetrazolium bromide) and Nitro BT (2,2'-di-p-nitrophenyl-3,3'-(3,3'-dimethoxy-4,4'-biphenylene) ditetrazolium chloride) have formed the basis of many techniques for the detection of dehydrogenases and enzymes which can be coupled to dehydrogenases. This method of staining makes use of the fact that the protons transferred to NAD or NADP to form $NADH_2$ or $NADPH_2$ may be taken up by phenazine methosulphate and thence transferred to a tetrazolium salt, which is thereby reduced to an insoluble formazan, which usually appears as a purple band.

Staining for dehydrogenases and other enzymes using phenazine methosulphate and a tetrazolium salt must be carried out in the dark, otherwise the background becomes stained because of light sensitivity of these reagents.

The following incubation medium has proved very satisfactory. It contains 10 ml 0·3 M Tris–HCl buffer pH 7·4, 5 ml 0·33 M sodium lactate, 10 mg NAD, 5 mg KCN, 3 mg MTT, 0·25 ml phenazine methosulphate (1 mg/ml). The gel slice is covered with this and incubated for up to 30 minutes at 37°C.

After staining, the gel slice is washed with frequent changes of distilled water for 30 minutes and then covered for storage with 80–90% glycerol solution which renders the gel slice tough and transparent.

Allen J. M. (1961) employs a similar staining technique but makes use of a variety of closely related substrates and coenzyme analogues.

A method for selective display of lactate dehydrogenase isoenzymes by incorporating an inhibitor into the incubation medium has recently been developed. With excess lactate (0·72 M) LDH-1 is not detected and with urea (2·6 M) LDH-5 is inhibited (Brody, 1964).

Quantitative evaluation of LDH isoenzyme patterns after electrophoresis. Enzymoelectrophoresis, as described by Wieme (1959a) is a sensitive method for the quantitative determination of LDH isoenzyme patterns using ultraviolet light.

Following visual staining of the serum LDH isoenzyme pattern after agar-gel electrophoresis, Van der Helm *et al.* (1962) have used reflectance densitometry of the isoenzyme bands and reported a linear relationship between peak area and enzyme activity with serum samples of 2–10 μl. Starkweather *et al.* (1966a) have described variations in the colours of the formazans produced in agar gels by reduction of nitroblue tetrazolium and suggest this is due to the production of monoformazans, as shown by Pearse (1957). Careful recrystallization of some of the commercially available tetrazolium salts, such as Nitro BT, is recommended, otherwise densitometry may be affected (Starkweather *et al.*, 1966a). Other methodological observations made by this group of workers concerned sample application to the agar gel. They found that if the sample is placed in a slot in the agar gel electrodecantation occurs and the zones of enzyme activity which are subsequently visualized are found to be located in the bottom half of the gel. They have recommended an application technique in which the sample is allowed to diffuse into the upper surface of the gel, using a disc with a centre slot to position the sample on the gel surface. Allison *et al.* (1963) have used a microscope with photometer attachment to quantitate the LDH isoenzymes in starch gels after tetrazolium salt staining and Preston *et al.* (1965) have obtained satisfactory quantitation of the LDH isoenzymes with visual staining following cellulose acetate electrophoresis.

Mull and Starkweather (1965) have used a microscope with a mechanical stage connected up to a conventional electrophoretic strip densitometer (Spinco model RA Analytrol) for quantitative determination of the LDH isoenzymes and claim reproducible results.

Reflectance densitometry has been used to provide reliable results for quantitative LDH isoenzyme distributions following starch gel electrophoresis of human tissue extracts (Latner and Turner, 1967). Peak area seems to be a better parameter than peak height and the only limitation of the method is that the response of the instrument is linear only below an individual isoenzyme concentration of 400 I.U./l.

Gotts and Skendzel (1966) have made a comparative study of

electrophoresis in agarose gel, followed by tetrazolium salt staining with subsequent elution of the formazan and the batch technique of adsorption-elution on DEAE-Sephadex. They have shown that the batch-type of adsorption-elution technique is unreliable when compared with electrophoretic analysis. With the former technique, there was apparently incomplete adsorption and elution of the isoenzymes on the DEAE-Sephadex.

(ii) *Malate Dehydrogenase*. The methods for demonstrating this group of isoenzymes have been ultraviolet light illumination and direct staining.

Ultraviolet Light Technique. The enzymoelectrophoresis technique has been employed by Lowenthal *et al.* (1961a) using oxaloacetate as substrate instead of pyruvate.

Visual Staining Technique. The methods first described (Markert and Møller, 1959; Tsao, 1960) have been improved by Thorne *et al.* (1963) who, after starch-gel electrophoresis at pH 8·6, used an incubation medium consisting of 23 ml 0·1 M Tris–HCl pH 8·5, 1·5 ml 2 M malate, 18 mg NAD, 0·6 ml phenazine methosulphate (1 mg/ml), 10 mg Nitro BT. The gel slice was incubated in this medium for 15–20 minutes at 37°C.

(iii) *Isocitrate Dehydrogenase*. A visual detection method has been employed after electrophoresis at pH 8·6 (Baron and Bell, 1962). Electrophoresis is probably better carried out at pH 6·2 due to the instability of isocitrate dehydrogenase above pH 7·8 (Campbell and Moss, 1962). An improvement on these techniques has been achieved by using starch-gel electrophoresis at pH 7·0 and pH 6·0 (Henderson, 1965). After electrophoresis the gel strips are incubated at 37°C for $\frac{1}{2}$–1 hour in a medium consisting of 45 ml 0·2 M Tris–HCl pH 8·0, 5 ml Nitro BT tetrazolium (1 mg/ml), 5 ml phenazine methosulphate (1·6 mg/ml), 3 ml 0·1 M sodium isocitrate, 1 ml NADP (10 mg/ml) and 0·2 ml manganese chloride (0·2 ml).

(iv) *Glutamate Dehydrogenase*. Attempts to fractionate glutamate dehydrogenase by starch-gel electrophoresis have been unsuccessful owing to the limiting pore size of the gel (Markert and Møller, 1959; Tsao, 1960). After agar-gel electrophoresis at pH 8·6 of human tissue extracts, up to five zones of activity can be detected with a substrate consisting of 1 ml M sodium DL-glutamate, 10 mg NAD, 1 ml 0·1 M KCN, 2·5 ml 0·5 M phosphate buffer pH 7·4, 2·5 mg Nitro BT and 0·25 ml phenazine methosulphate solution (1 mg/ml) (Van der Helm, 1962b). Incubation is carried out at 37°C for 1 hour.

(v) *Glucose-6-phosphate Dehydrogenase.* Several methods have been reported for the demonstration of glucose-6-phosphate dehydrogenase after starch-gel electrophoresis of red cell haemolysates (Marks *et al.*, 1961; Kirkman, 1962; Boyer *et al.*, 1962). After starch-gel electrophoresis at pH 8·8 with 0·05 M Tris–HCl–2·7 mM EDTA as both bridge and gel buffer, and with the cathode buffer compartment and the gel buffer being 10 μM with respect to NADP, Kirkman and Hendrickson (1963) have used an incubation medium for visualizing the enzyme which involves methylene blue and Nitro BT. An incubation medium consisting of 13 ml 0·3 MTris–HCl buffer pH 7·6, 5 mg NADP, 1 ml 0·025 M glucose-6-phosphate, 1 ml 0·1 M MgCl$_2$, 0·25 ml phenazine methosulphate (1 mg/ml) and 3 mg MTT is used in the authors' laboratory. The gel slice is incubated in this for 30 minutes at 37°C.

(vi) *Phosphogluconate Dehydrogenase.* This enzyme has been demonstrated after starch-gel electrophoresis of human red cell haemolysates using phosphate buffers at pH 7·0 (Fildes and Parr, 1963). Electrophoresis is followed by incubation in a medium containing 10 ml 0·1 M Tris–HCl buffer pH 8·6, 100 mg agar, 2 mg NADP, 10 mg sodium phosphogluconate, 0·4 mg phenazine methosulphate and 2 mg MTT. The gel slice is incubated for about 1 hour at 37°C. In our laboratory phosphogluconate dehydrogenase isoenzymes in extracts of human tumour tissues have been rendered visible after starch-gel electrophoresis, using a medium containing 13 ml 0·3 M Tris–HCl buffer pH 8·6, 5 mg NADP, 1 ml 25 mM sodium phosphogluconate, 0·25 mg phenazine methosulphate and 2 mg MTT. The gel slice is incubated in this medium for 30 minutes at 37°C.

(vii) *Galactose-6-phosphate Dehydrogenase.* Some of the multiple forms of glucose-6-phosphate dehydrogenase in human, horse and deermouse tissues have been found to be equally active with galactose-6-phosphate as substrate. Starch-gel electrophoresis using 0·05M Tris–EDTA–sodium borate buffer pH 8·0 can be used to separate these isoenzymes and the zones of enzyme activity are visualized by incubating the gel slice for up to 4 hours at 35°C in 100 ml of staining solution containing 50 mg Nitro-blue tetrazolium, 50 mg NADP, 2 mg phenazine methosulphate, 10 ml Tris buffer pH 6·8 and 5 ml M galactose-6-phosphate (Shaw, 1966b).

(viii) *Alcohol Dehydrogenase.* Agar-gel electrophoresis of homogenates of *Drosophila* followed by direct staining for enzyme activity has been used to demonstrate up to seven components of this enzyme (Ursprung and Leone, 1965). The staining medium employed contains 37·5 ml 0·5 M Tris buffer pH 9·0, 1·5 ml 95% ethanol, 0·375 ml phenazine methosulphate (2 mg/ml), 3·75 ml Nitro-blue tetrazolium (5 mg/ml) and 1·5 ml NAD (50 mg/ml). The gel is incubated in this medium for 15–30 minutes at 37°C.

B. TRANSFERASES

(i) *Aspartate Aminotransferase.* This enzyme has been detected by means of ultraviolet light and by direct staining.

Boyd (1961), using agar gel electrophoresis, and Boyde and Latner (1962), using starch-gel electrophoresis, have made use of ultraviolet light for detecting the bands of enzyme activity. The latter workers employ an incubation medium consisting of 1·5 ml L-aspartate (0·05 M in 0·1 M phosphate buffer pH 7·4), 1·5 ml α-oxoglutarate (0·005 M in 0·1 M phosphate buffer pH 7·4), 0·2 ml pyridoxal phosphate (500 μg/ml), 0·05 ml malate dehydrogenase (0·5 mg protein/ml Boehringer), 3 mg $NADH_2$ and 1·5 ml water. The gel is incubated at 37°C for up to 4 hours. The areas of enzyme activity appear as light bands on a dark background when viewed by ultraviolet light. The principle of the method depends on coupling the transferase with malate dehydrogenase and thereby detecting regions where $NADH_2$ has been diminished by the conversion of oxaloacetate to malate.

This ultraviolet light detection method can be converted to a direct staining method by adding a solution of 4 mg MTT and 0·3 mg phenazine methosulphate in 15 ml water to the gel slice after incubation in the substrate. The zones of transferase activity show up as pale areas on a purple background.

Another method for staining aspartate aminotransferases after starch-gel electrophoresis has been described which employs a diazonium salt Azoene Fast Violet B which complexes specifically with oxaloacetate (Decker and Rau, 1963; Schwartz *et al.*, 1963). The incubation medium consists of 5·3 ml 0·2 M phosphate buffer pH 7·4, 0·75 g polyvinyl-pyrrolidine (PVP), 0·2 ml pyridoxal-5-phosphate (500 μg/ml), 0·4 ml bovine albumin (30 mg/ml), 1·7 ml 0·2 M aspartate, 0·7 ml 0·1 M α-oxoglutarate, 50 mg Azoene Fast Violet B (6-Benzamide-4-methoxy-m-toluidine diazonium chloride) and 1·7 ml water (Decker and Ran, 1963). The gel slice is incubated in this medium for up to 1 hour at 37°C.

(ii) *Hexokinase.* Direct staining for enzyme activity following starch-gel electrophoresis in 0·02 M barbitol buffer containing 5 mM mercapto-ethanol—2·7 mM EDTA (pH 8·4) has been achieved (Katzen and

Schimke, 1965). The gel slice is incubated at 37°C for 90 minutes in a medium containing the following in the final concentrations stated: 0·1 M Tris buffer pH 7·4, 5 mM NADP, 5 mM $MgCl_2$, 2 mM KCN, 5 mM ATP, 0·5 mM or 1·0 mM glucose, 0·4 I.U./ml glucose-6-phosphate dehydrogenase, 40 μg/ml phenazine methosulphate and 400 μg/ml Nitro BT.

(iii) *Creatine Kinase.* Creatine kinase activity can be detected after agar-gel enzymoelectrophoresis at pH 9·0 by means of ultraviolet light. The 3% agar mixed with 3 ml 433 mM glycine buffer pH 9·0 containing the following at the final concentrations stated: 3 mM magnesium acetate, 68 mM creatine, 3·5 mM ATP, 1·2 mM phosphoenolpyruvate and 1·4 mM $NADH_2$. The mixture is heated until the agar dissolves, the solution cooled to 50°C and 0·125 mg/ml lactate dehydrogenase and 0·025 mg/ml pyruvate kinase added. The resultant solution is poured on to a microscope slide, allowed to set and placed on the electrophoresis gel as is usual for enzymoelectrophoresis. The zones of enzyme activity show up under ultraviolet light after incubation at 30°C for 10–30 minutes. Pyruvate kinase can be demonstrated by omitting creatine and pyruvate kinase from the incubation medium (Fellenberg *et al.*, 1963).

Creatine kinase activity can be detected on cellulose acetate strips after electrophoresis at pH 8·6 in barbitone buffer (Rosalki, 1965). The staining medium consisting of 2 mg ADP, 10 μl hexokinase suspension (2·8 I.U.), 10 μl glucose-6-phosphate dehydrogenase suspension (1·4 I.U.), 2 mg glucose, 7 mg $MgSO_4.7H_2O$, 3 mg NADP, 1·6 mg MTT, 3 ml Tris buffer (0·05 M) pH 7·5, 80μl phenazine methosulphate (1 mg/ml). The strip is incubated in this medium for 1 hour at 37°C.

(iv) *Phosphoglucomutase.* Multiple forms of this enzyme can be demonstrated following starch-gel electrophoresis in 0·01 M Tris–maleate–EDTA buffer pH 7·4 being 0·001 M in $MgCl_2$ (Spencer *et al.*, 1964). The gel slice is incubated at 37°C for 1 hour in a medium containing the following at the final concentration stated: 4·6 mM glucose-1-phosphate, 50 μM glucose-1-6-diphosphate, 10 mM $MgCl_2$, 0·04 I.U./ml glucose-6-phosphate dehydrogenase, 0·1 mg/ml phenazine methosulphate, 0·1 mg/ml MTT tetrazolium salt, made up in 0·03 M Tris buffer pH 8·0. Sufficient amounts of glucose-1-6-diphosphate often occur in commercial preparations of glucose-1-phosphate so that additional glucose-1-6-diphosphate is not required.

C. HYDROLASES

(i) *Alkaline Phosphatase*

Extraction Methods. For some studies of alkaline phosphatases (Kowlessar *et al.*, 1958; 1961; Hodson *et al.*, 1962) the gel segments have been macerated directly with the buffered substrate before assay of the enzyme activity and the mixture as for the normal colorimetric assay procedure incubated. After centrifugation, the colour development in enzyme activity and the mixture incubated as for the normal colorimetric assay procedure. After centrifugation, the colour development in the supernatant has been determined.

Ultraviolet Light Techniques. After starch-gel electrophoresis, strips of filter paper moistened with a solution of disodium α-naphthyl phosphate (5 mM) in bicarbonate buffer at pH 10 are placed on the gel slice. After incubation for 5–10 minutes at 37°C, the papers are removed and the gel viewed under ultraviolet light. Owing to liberation of α-naphthol, enzymatically active areas show as bands of pale blue fluorescence on a dark fluorescent background (Moss *et al.*, 1961a).

Visual Staining Techniques. Visual demonstration after starch-gel electrophoresis was first achieved by Estborn (1959) who used test papers containing α-naphthyl phosphoric acid and diazo-*o*-dianisidin, buffered appropriately. After incubation, coloured bands appeared on the paper. Actual staining of activity in the gel was later developed (Allen and Hunter, 1960; Dubbs *et al.*, 1960; Boyer, 1961; Hodson *et al.*, 1961; Stevenson, 1961). These also all depend on the coupling of naphthol liberated from α-naphthyl phosphate with diazonium salts, such as Fast Blue RR (Boyer, 1961) and Brentamine Fast Red TR (Hodson *et al.*, 1962). Although many workers recommend diazonium salt concentrations of up to 50 mg % or over, it has been found that concentrations of 5–10 mg % give best results. For demonstration of human tissue and serum alkaline phosphatases after starch-gel electrophoresis using 0·05 M Tris–HCl buffers pH 8·8, the recommended staining medium consists of 10 mg calcium α-naphthyl phosphate, and 5 mg Brentamine Fast Red TR dissolved in 100 ml 1·0 mM $MgCl_2$/0·1 M borax pH 9·6. The gel slice is incubated in 50 ml of this buffered substrate at 37°C for 1 hour with three changes of the substrate (Hodson *et al.*, 1962). These changes of the substrate are necessary to remove breakdown products of the reaction and also to maintain the concentration of the substrate in contact with the gel. After incubation, the staining solution is poured off and replaced by glycerine, which makes the gel transparent. The zones of enzyme activity appear as reddish brown areas. This type of technique, where low concentrations of diazonium salt are used, has been found in our laboratory to give the best results, not only with phosphatases but also with other hydrolases. If higher concentrations are used, non-specific staining of high density protein zones such as albumin occurs and in some cases the diazonium salts are known to be inhibitory to the enzymes.

Specific alkaline phosphatases can also be demonstrated after gel electrophoresis (Sandler and Bourne, 1961; Allen and Hynick, 1963). The method employing lead phosphate (Allen and Hynick, 1963) seems the more satisfactory, as it compares favourably with the azo-coupling methods and is twenty times as sensitive as the alizarin method (Sandler and Bourne, 1961). The technique involves the conversion of calcium phosphate precipitates, formed by enzymic activity, into lead phosphate by immersion in buffered lead nitrate. After washing and immersion in ammonium sulphide, the sites of enzymic activity are visualized. The gel slice is incubated in 0·05 M sodium β-glycerophosphate or other appropriate phosphate ester in a 15 mM calcium chloride, 33 mM Tris solution adjusted to pH 9·5. After 15 minutes incubation at 37°C, this solution is replaced by 3 mM lead nitrate in 0·08 M Tris–maleate pH 7·0. After a further 30 minutes at room temperature, the lead nitrate solution is decanted and the gel washed for 1 hour with frequent changes of distilled water. The zones of phosphatase activity are then rendered visible by treating the gel slice with 5% ammonium sulphide solution for 2 minutes.

(ii) *Acid Phosphatase.* Although the earlier techniques employed starch-gel electrophoresis at pH 8·6, it was soon realized that the enzyme was inhibited at this pH. A subsequent gel buffer recommended for this enzyme was 0·005 M Tris–citrate pH 6·2 (Sur *et al.*, 1962).

Acid phosphatases were first demonstrated after starch-gel electrophoresis (Estborn, 1959; 1961) using a test paper impregnated with α-naphthyl phosphoric acid and diazo-*o*-dianisidin and moistened with acetate buffer pH 5·0. The test paper was placed on the gel slice for up to 2 hours at 37°C.

Several techniques have been used after acrylamide-gel electrophoresis of rat liver phosphatases (Barka, 1961). The first of these employed α-glycerophosphate in a similar manner to the standard Gomori (1950) technique. A post-coupling method makes use of sodium 6-benzoyl-2-naphthol phosphate for incubation with subsequent addition of Fast Blue B. Simultaneous azo-dye coupling techniques with α-naphthyl phosphate as substrate and Fast Garnet GBC or hexazonium pararosaniline as couplers or Naphthol AS-BI phosphate as substrate with Fast Red Violet LB as coupler have been compared and the latter proved more satisfactory.

The genetically determined red cell acid phosphatases cannot apparently be demonstrated by azo-dye techniques and the method employed for their demonstration involves the use of phenolphthalein phosphate as substrate (Hopkinson *et al.*, 1963; 1964). The gel slice is incubated for 3 hours at 37°C in 50 ml 0·005 M phenolphthalein phosphate in 0·05 M citrate buffer at pH 6·0. The buffered substrate is then decanted and the acid phosphatases visualized as red zones by pouring 2 ml of concentrated ammonia solution over the gel slice.

(iii) *Esterases*. Following the development of the "zymogram" technique for esterases (Hunter and Markert, 1957), three basic techniques have been elaborated for demonstration of esterase activity after gel electrophoresis (Hunter and Burstone, 1960).

The first employs a diazonium salt for coupling with α- or β-naphthol liberated from an appropriate ester. A whole range of esters have been employed varying from α- or β-naphthyl acetate to α-naphthyl cinnamate or β-naphthyl chloroacetate. The diazonium salt generally used is Fast Blue RR, although Fast Garnet GBC, Fast Blue B or Fast Blue BB are satisfactory. The incubation medium employed consists of 2 ml of a 1% solution of the appropriate substrate in acetone, 2 ml 0·2 M Tris buffer pH 7·0, 47 ml distilled water and 25 mg of the diazonium salt.

The second technique makes use of either of the two substrates indoxyl acetate or O-acetyl-5-bromoindoxyl. A blue colour is produced from the liberated indoxyl or its derivative by means of a mixture of ferrocyanide and ferricyanide. The incubation medium consists of 2 ml ethoxyethanol containing 10 mg substrate, 15 ml 0·2 M Tris buffer pH 7·0, 5 ml 0·05 M ferrocyanide, 5 ml 0·05 M ferricyanide and 15 ml distilled water.

The third technique employs Naphthol AS esters and diazonium salts. The incubation medium consists of 1 ml acetone containing 10 mg substrate, 10 ml ethoxyethanol, 10 ml 0·2 M Tris buffer pH 7·0, 29 ml distilled water and 25 mg diazonium salt.

For each of the above three methods incubation is carried out at 37°C for periods of time ranging from 30 minutes to 3 hours.

In studies of substrate specificity of esterases, the sensitivity to inhibitors and activators is often employed as well as specific substrates. For inhibition studies the gels are placed in the inhibitor solution for 10–15 minutes prior to addition of the substrate. The same inhibitor is also incorporated within the substrate medium. Sodium taurocholate is the most common activator used and the inhibitors include eserine, di-isopropyl fluorophosphate (DFP) and tetraethyl pyrophosphate

(TEPP). Although the use of inhibitors has been used to demonstrate cholinesterase activity, it can also be demonstrated by specific substrates, such as β-carbonaphthoxy choline iodide (Dubbs *et al.*, 1960) and 6-bromo-2-naphthyl carbonaphthoxy choline iodide (Lawrence *et al.*, 1960). The staining medium for the latter consists of 20 mg substrate and 25 mg Fast Blue B dissolved in 50 ml 0·1 M Tris maleate buffer pH 6·4, the incubation being carried out at room temperature for 1 hour.

(iv) *Arylamidase or Leucine Aminopeptidase.* Although activity has been shown by homogenization of the gel segments in an appropriate substrate (Kowlessar *et al.*, 1960; 1961), the enzyme is best demonstrated after gel electrophoresis by diazonium salt coupling following action on a substrate such as leucyl- or alanyl-β-naphthylamide (Dubbs *et al.*, 1960; 1961). A suitable staining medium for use after starch-gel electrophoresis at pH 8·6 in Tris–HCl or borate buffers consists of 10 ml 0·1 M acetate buffer pH 6·5, 5 mg DL-alanyl-β-naphthylamide and 3 mg Brentamine Fast Red TR. Incubation is carried out at 37°C for 30 minutes with two changes of substrate.

(v) *Peptide Peptidohydrolases.* Proteolytic enzyme activity has been detected after agar-gel electrophoresis of tissue extracts and purified enzyme preparations (Uriel, 1960). After electrophoresis the plate is immersed in an appropriate protein substrate and after incubation the plate is fixed and stained for protein in the normal manner. The zones of enzyme activity show up as unstained areas on the gel. Trypsin has also been demonstrated after starch-gel electrophoresis using *N*-benzoyl-*dl*-arginine-β-naphthylamide and Fast Blue BB (Hess *et al.*, 1963).

(vi) *β-glucuronidase.* This enzyme can be detected after starch-gel electrophoresis by means of 8-hydroxyquinoline glucuronide which liberates 8-hydroxyquinoline to couple with a diazonium salt Fast Blue RR (Lawrence *et al.*, 1960). The incubation medium consists of 15 mg 8-hydroxyquinoline glucuronide, 20 mg Fast Blue RR dissolved in 30 ml 0·1 M acetate buffer pH 5·2. The gel is incubated for 16 hours at 37°C.

(vii) *Amylase.* Paper electrophoresis of human serum, followed by elution of the fractions and amyloclastic (starch-iodine reaction) determination of amylase activity does not give satisfactory results due to interaction of some of the serum protein fractions with iodine (Ujihira *et al.*, 1965; Searcy *et al.*, 1965). A saccharogenic method involving incubation

I

at 37°C for 60 minutes of the eluted protein fractions with 1% soluble starch buffered with 0·1 M phosphate buffer pH 6·9, followed by colour development with a 1% 3,5-dinitrosalicylic reagent has been recommended (Ujihira *et al.*, 1965).

Agar-gel electrophoresis of serum body fluids and tissue extracts does, however, yield results with visual staining by an amyloclastic technique which show the non-identity of pancreatic and salivary amylases (Oger and Bischops, 1966). The agar electrophoresis plate is placed in a solution containing 1% soluble starch and 1% agar. After incubation at 37°C for periods from 20 minutes to 4 hours, the plates are fixed and dried and the zones of enzyme activity visualized with 0·01 M iodine solution.

3. IMMUNOLOGICAL TECHNIQUES

The use of specific characterization reactions in immunodiffusion and immunoelectrophoretic studies has shown that enzymes may be examined using such techniques with reference to (a) the identification of an enzyme within a mixture of enzymes and proteins, (b) the study of purity and homogeneity of enzymes and (c) the immunological relationships between isoenzymes (Uriel, 1963). These techniques have provided data on peroxidases, catalases, caeruloplasmin, carboxylic esterases and some pancreatic proteases. Techniques of immunoelectrophoresis have been used in studies of human serum esterases (Uriel, 1961) adult and embryonic chicken serum esterases (Croisille, 1962) and mouse serum esterases (Hermann *et al.*, 1963). Immunoelectrophoresis has also been used in a study of the enzymatic activity related to human serum β-lipoprotein: the enzymes studied included dehydrogenases, phosphatases, esterases, aminopeptidases and β-glucuronidase (Lawrence and Melnick, 1961). Various dehydrogenases have been examined but the results have been difficult to interpret (Brown and Avrameas, 1963).

A method for immunoelectrophoretic studies of proteolytic enzymes (Raunio and Gabriel, 1963) is of special interest as it may be applicable to studies of other types of enzymes. After agar immunoelectrophoresis, a second agar plate containing 1% haemoglobin in 0·1 M glycine–HCl buffer, pH 2·0, is placed on top of the immunoelectrophoresis plate as in the enzymoelectrophoresis technique. The slide "sandwich" is incubated for 1 hour at 37°C and then stained. A transparent area the same shape as the immunodiffusion arc indicates the presence of the proteolytic activity. The immunoelectrophoresis plate may then be re-used for another type of enzyme. Attempts at direct staining for enzyme activity can often be obscured by the semi-opaque nature of the precipitin lines. In this respect the "sandwich" method has a great advantage.

An interesting immunological technique has been developed for the study of specific isoenzyme antigens in tissue sections (Nace *et al.*, 1961; Nace, 1963). This technique has been applied mainly to lactate dehydrogenase isoenzymes and involves agar immunoelectrophoresis of an extract of the tissue in question. After immunoelectrophoresis and development of the precipitin pattern using a fluorescent labelled antiserum, the agar plate is sliced and one of the slices stained for enzyme activity in the usual manner. The antigens to be studied are thus localized and the corresponding precipitin lines from the unstained agar immunoelectrophoresis plate are cut out and minced with an equal volume of normal saline. Uncombined and non-specific reagents are removed by washing the agar gel mince with two to three changes of normal saline for two hours. The washed, minced agar containing the specific antigen-antibody precipitate is placed on the tissue sections. The whole is adjusted to pH 10·5 by covering with a small amount of 0·1 M borate–carbonate buffer. After diffusion at 22°C for 2 hours, during which the fluorescent antibody diffuses into the tissue section, a small amount of 0·14 M HCl is added and the antibody now combines with the specific enzyme antigen in the tissue section. The section is then washed with distilled water for up to 1 hour to remove excess antigen, antibody, antigen–antibody precipitates and agar. Standard fluorescence microscopy is employed for visualization of the location of the appropriate isoenzyme.

References

Aalund, O., Rendel, J. and Freedland, R. A. (1965). *Biochim. Biophys. Acta* **110**, 113.
Abdul-Fadl, M. A. M. and King, E. J. (1947). *Biochem. J.* **41**, xxxii.
Abdul-Fadl, M. A. M. and King, E. J. (1948). *Biochem. J.* **42**, xxviii.
Abdul-Fadl, M. A. M. and King, E. J. (1949). *Biochem. J.* **45**, 51.
Acheson, J., James, D. C., Hutchinson, E. C. and Westhead, R. (1965). *Lancet* **i**, 1306.
Adams, E. and Finnegan, C. V. (1965). *J. exp. Zool.* **158**, 241.
Agnall, I. P. S. and Kjellberg, B. (1965). *Comp. Biochem. Physiol.* **16**, 512.
Agostini, A., Vergana, C. and Villa, L. (1966). *Nature, Lond.* **209**, 1024.
Aldridge, W. N. (1953a). *Biochem. J.* **53**, 110.
Aldridge, W. N. (1953b). *Biochem. J.* **53**, 110.
Aldridge, W. N. (1954). *Biochem. J.* **57**, 692.
Allen, J. M. (1961). *Ann. N.Y. Acad. Sci.* **94**, 937.
Allen, J. M. (1963). *J. Histochem. Cytochem.* **11**, 542.
Allen, J. M. and Hunter, R. L. (1960). *J. Histochem. Cytochem.* **8**, 50.
Allen, J. M. and Hynick, G. (1963). *J. Histochem. Cytochem.* **11**, 169.
Allen, S. L. (1960). *Genetics, Princeton* **45**, 1051.
Allen, S. L. (1961). *Ann. N.Y. Acad. Sci.* **94**, 753.
Allen, S. L. (1964). *J. exp. Zool.* **155**, 349.
Allen, S. L., Misch, M. S. and Morrison, B. M. (1963a). *Genetics, Princeton* **48**, 1635.
Allen, S. L., Misch, M. S. and Morrison, B. M. (1963b). *J. Histochem. Cytochem.* **11**, 706.
Allen, S. L., Allen, J. M. and Licht, B. M. (1965). *J. Histochem. Cytochem.* **13**, 434.
Alles, G. A. and Hawes, R. C. (1940). *J. biol. Chem.* **133**, 375.
Allison, M. J., Gerszten, E. and Sanchez, M. (1963). *Am. J. clin. Path.* **40**, 451.
Allison, M. J., Gerszten, E. and Sanchez, B. (1964). *Endocrinology* **74**, 87.
Allott, E. N. and Thompson, J. C. (1956). *Lancet, Lond.* **ii**, 517.
Amelung, D. I. (1963). *Dtsch. med. Wschr.* **88**, 1940.
Anderson, B. M., Ciotti, C. J. M. and Kaplan, N. O. (1959). *J. biol. Chem.* **234**, 1219.
Andrews, P. (1962). *Nature, Lond.* **196**, 36.
Angeletti, P. U., Suntzeff, V. and Moore, B. W. (1960a). *Cancer Res.* **20**, 1229.
Angeletti, P. U., Moore, B. W. and Suntzeff, V. (1960b). *Cancer Res.* **20**, 1592.
Anstall, H. B., Lapp, C. and Trujillo, J. M. (1966). *Science, N.Y.* **154**, 657.
Appel, S. H., Alpers, D. H. and Tomkins, G. M. (1965). *J. Molec. Biol.* **11**, 12.
Appella, E. and Markert, C. L. (1961). *Biochem. biophys. Res. Commun.* **6**, 171.
Arfors, K. E., Beckman, L. and Lundin, L. G. (1963a). *Acta genet. Statist. Med.* **13**, 89.
Arfors, K. E., Beckman, L. and Lundin, L. G. (1963b). *Acta genet. Statist. Med.* **13**, 366.
Arfshapour, F. and O'Brien, R. D. (1963). *J. Insect Physiol.* **9**, 521.
Armstrong, A. R. and Banting, F. G. (1935). *Canad. med. Ass. J.* **33**, 243.
Augustinsson, K–B. (1958). *Nature, Lond.* **181**, 1786.
Augustinsson, K–B. (1959). *Acta. chem. scand.* **13**, 571, 1081, 1097.
Augustinsson, K-B. (1961). *Ann. N.Y. Acad. Sci.* **94**, 844.
Augustinsson, K-B. and Erne, K. (1961). *Experimentia* **17**, 396.
Avrameas, S. and Rajewsky, K. (1964). *Nature, Lond.* **201**, 405.
Aw, S. E. (1966). *Nature, Lond.* **209**, 298.
Bach, M. L., Singer, E. R., Levinthal, C. and Sizer, I. W. (1961). *Fedn. Proc. Fedn. Am. Socs. exp. Biol.* **20**, 255.

Baker, R. W. R. and Pellegrino, C. (1954). *Scand. J. clin. Lab. Invest.* **6**, 94.
Bamford, K. F. and Harris, H. (1964). *Ann. hum. Genet.* **27**, 417.
Bamford, K. F., Harris, H., Luffman, J. E., Robson, E. B. and Cleghorn, T. (1965). *Lancet* **i**, 530.
Bar, U., Schmidt, E. and Schmidt, F. W. (1963). *Klin. Wschr.* **41**, 977.
Barka, T. (1961). *J. Histochem. Cytochem.* **9**, 564.
Barnett, H. (1962). *Lancet* **ii**, 199.
Barnett, H. (1964). *J. clin. Path.* **17**, 567.
Barnett, H. and Gibson, A. (1964). *Proc. Ass. clin. Biochem.* **3**, 6.
Baron, D. N. and Bell, J. L. (1962). *Proc. Ass. clin. Biochem.* **2**, 8.
Barron, K. D. and Bernsohn, J. (1965). *Ann. N.Y. Acad. Sci.* **122**, 369.
Barron, K. D., Bernsohn, J. I. and Hess, A. R. (1961). *J. Histochem. Cytochem.* **9**, 656.
Barron, K. D., Bernsohn, J. I. and Hess, A. R. (1963). *J. Histochem. Cytochem.* **11**, 139.
Barron, K. D., Bernsohn, J. and Hess, A. R. (1964). *J. Histochem. Cytochem.* **12**, 42.
Barron, K. D., Bernsohn, J. and Hess, A. R. (1966). *J. Histochem. Cytochem.* **14**, 1.
Batsakis, J. G., Preston, J. A. and Briere, R. O. (1964). *Milit. Med.* **129**, 1161.
Baume, P. E., Builder, J. E., Fenton, B. H., Irving, L. G. and Piper, D. W. (1966). *Gastroenterology* **50**, 781.
Baumgarten, A. (1963). *Blood* **22**, 466.
Beckman, L. (1964). *Acta genet. Statist. Med.* **14**, 286.
Beckman, L. (1966). "Isozyme variations in man". Karger, Basel and New York.
Beckman, L. and Grivea, M. (1965). *Acta genet. Statist. Med.* **15**, 218.
Beckman, L. and Johnson, F. M. (1964a). *Hereditas* **51**, 221.
Beckman, L. and Johnson, F. M. (1964b). *Genetics, Princeton* **49**, 829.
Beckman, L. and Johnson, F. M. (1964c). *Hereditas* **51**, 212.
Beckman, L. and Regan, J. D. (1964). *Acta path. microbiol. scand.* **62**, 567.
Beckman, L., Bjorling, G. and Christodoulou, C. (1966). *Acta genet. Statist. Med.* **16**, 59.
Beckman, L., Scandalios, J. and Brewbacker, J. L. (1964a). *Genetics, Princeton* **50**, 899.
Beckman, L., Scandalios, J. G. and Brewbacker, J. L. (1964b). *Science, N.Y.* **146**, 1174.
Bell, J. L. and Baron, D. N. (1964). *Biochem. J.* **90**, 8P.
Bell, R. L. (1963). *Am. J. clin. Path.* **40**, 216.
Bergkvist, R. (1963). *Acta chem. scand.* **17**, 1541.
Berk, J. E., Kawaguchi, M., Zeinch, R., Ujihira, I. and Searcy, R. (1963). *Nature, Lond.* **200**, 572.
Berk, J. E., Searcy, R. L., Hayashi, S. and Ujihira, I. (1965). *J. Am. med. Ass.* **192**, 389.
Bernsohn, J., Barron, K. D. and Hess, A. R. (1961). *Proc. Soc. exp. Biol. Med.* **108**, 71.
Bernsohn, J. I., Barron, K. D., Hess, A. R. and Hedrick, T. (1963). *J. Neurochem.* **10**, 783.
Bessman, S. P. and Fonyo, A. (1966). *Biochem. biophys. Res. Commun.* **22**, 597.
Beutler, E. and Collins, Z. (1965). *Science, N.Y.* **150**, 1306.
Bierman, H. R., Hill, B. R., Reinhardt, L. and Emery, E. (1957). *Cancer Res.* **17**, 660.
Biron, P. (1964). *Revue can. Biol.* **23**, 497.
Blanchaer, M. C. (1961a). *Clinica chim. Acta* **6**, 272.
Blanchaer, M C. (1961b). *Pure appl. Chem.* **3**, 403.
Blanchaer, M. C. and Van Wijhe, M. (1962). *Nature, Lond.* **193**, 877.
Blanchaer, M. C. and Van Wijhe, M. (1962). *Am. J. Physiol.* **202**, 827.
Blanco, A. and Zinkham, W. H. (1963). *Science, Lond.* **139**, 601.
Blanco, A. and Zinkham, W. H. (1966). *Bull. Johns Hopkins Hosp.* **118**, 27.
Blatt, W. F., Blatteis, C. M. and Mager, M. (1966). *Can. J. Biochem. Physiol.* **44**, 537.

Blostein, R. E. and Rutter, W. J. (1963). *J. biol. Chem.* **238**, 3280.
Bonovita, V. (1965). *In* "First Intern. Symp. Biochemistry of the Retina" London, 1964, p. 5, (C. N. Graymore, ed.) Academic Press, London.
Bonavita, V. and Guarneri, R. (1962). *Biochim. biophys. Acta*, **59**, 634.
Bonavita, V. and Guarneri, R. (1963a). *J. Neurochem.* **10**, 743.
Bonavita, V. and Guarneri, R. (1963b). *J. Neurochem.* **10**, 755.
Bonavita, V., Ponte, F. and Amore, G. (1962). *Nature, Lond.* **196**, 576.
Bonavita, V., Ponte, F. and Amore, G. (1963). *Vision Res.* **3**, 271.
Bonavita, V., Ponte, F. and Amore, G. (1964). *J. Neurochem.* **11**, 39.
Borst, P. and Peeters, E. M. (1961). *Biochim. biophys. Acta* **54**, 188.
Bottomley, R. H., Locke, S. J. and Ingram, H. C. (1964). *Proc. Am. Ass. Cancer Res.* **5**, 7.
Bouchilloux, S., McMahill, P. and Mason, H. S. (1963). *J. biol. Chem.* **238**, 1699.
Bourne, J. G., Collier, H. O. J. and Somers, G. F. (1952). *Lancet* **i**, 1225.
Bowman, J. E., Carson, P. E. and Frischer, H. (1966). *Nature, Lond.* **210**, 811.
Boyd, J. B. and Mitchell, H. K. (1965). *Analyt. Biochem.* **13**, 28.
Boyd, J. W. (1961). *Biochem. J.* **81**, 439.
Boyd, J. W. (1962). *Clinica chim. Acta* **7**, 424.
Boyd, J. W. (1966). *Biochim. Biophys. Acta* **113**, 302.
Boyde, T. R. C. and Latner, A. L. (1962). *Biochem. J.* **82**, 51P.
Boyer, S. H. (1961). *Science, N.Y.* **134**, 1002.
Boyer, S. H. (1963). *Ann. N.Y. Acad. Sci.* **103**, 938.
Boyer, S. H., Fainer, D. C. and Watson-Williams, E. J. (1963). *Science, N.Y.* **141**, 642.
Boyer, S. H., Porter, I. H. and Weilbacher, R. G. (1962). *Proc. natn. Acad. Sci. U.S.A.* **48**, 1868.
Brewer, G. J. and Dern, R. J. (1964). *Am. J. hum. Genet.* **16**, 472.
Brody, I. A. (1964). *Nature, Lond.* **201**, 685.
Brody, I. A. (1965). *Nature, Lond.* **205**, 196.
Brody, I. A. (1966). *J. Neurochem.* **13**, 975.
Brown, G. and Avrameas, S. (1963). *Bull. Soc. Chim. biol.* **45**, 233.
Brown, K. D. and Doy, C. H. (1966). *Biochim. Biophys. Acta* **118**, 157.
Bucher, T. and Klingenberg, M. (1958). *Angew. Chem. Ausg. B* **70**, 552.
Burger, A., Richterich, R. and Aebi, H. (1964). *Biochem. Z.* **339**, 305.
Burgi, von. H., Weismann, U. and Richterich, R. (1964). *Schweiz. med. Wnschr.* **94**, 1242.
Bush, G. H. (1961). *Br. J. Anaesth.* **33**, 454.
Buta, J. L., Conklin, J. L. and Davey, M. M. (1966). *J. Histochem. Cytochem.* **14**, 658.
Butler, E. A. and Flynn, F. V. (1961). *J. clin. Path.* **14**, 172.
Butterworth, P. J. and Moss, D. W. (1966). *Nature, Lond.* **209**, 805.
Butterworth, P. J., Moss, D. W., Pitkanen, E. and Pringle, A. (1965). *Clinica chim. Acta* **11**, 220.
Cahn, R. D. (1963). *J. Cell Biol.* **19**, 12A.
Cahn, R. D. (1964). *Dev. Biol.* **9**, 327.
Cahn, R. D., Kaplan, N. O., Levine, L. and Zwilling, E. (1962). *Science, Lond.* **136**, 962.
Campbell, D. M. and Moss, D. W. (1962). *Proc. Ass. clin. Biochem.* **2**, 10.
Cann, D. C. and Wilcox, M. E. (1965). *J. appl. Bact.* **28**, 165.
Carlsson, K., and Frick, G. (1964). *Biochim. biophys. Acta* **81**, 301.
Carr, A. and Skillen, A. W. (1963). *Br. J. Derm.* **75**, 331.
Carter, B. G., Cinader, B. and Ross, C. A. (1961). *Ann. N.Y. Acad. Sci.* **94**, 1004.
Cenciotti, L. and Mariotti, A. (1964). *Pathologica* **56**, 191.
Chiandussi, L., Greene, S. F. and Sherlock, S. (1962). *Clin. Sci.* **22**, 425.
Childs, V. A. and Legator, M. S. (1965). *Life Sci.* **4**, 1643.

Chilson, O. P., Costello, L. A. and Kaplan, N. O. (1964). *J. Molec. Biol.* **10**, 349.

Chilson, O. P., Costello, L. A. and Kaplan, N. O. (1965a). *Biochemistry, N.Y.* **4**, 271.

Chilson, O. P., Kitto, G. B. and Kaplan, N. O. (1965b). *Proc. natn. Acad. Sci. U.S.A.* **53**, 1006.

Chilson, O. P., Kitto, G. B., Pudles, J. and Kaplan, N. O. (1966). *J. biol. Chem.* **241**, 2431.

Chung, A. E. and Langdon, R. G. (1963). *J. biol. Chem.* **238**, 2309.

Chytil, F. (1965). *Biochem. biophys. Res. Commun.* **19**, 630.

Clausen, J. and Gerhardt, W. (1963). *Acta neurol. scand.* **39**, 305.

Clausen, J. and Ovlisen, B. (1965). *Biochem. J.* **97**, 513.

Coddington, A., Fincham, J. R. S. and Sundaram, T. K. (1966). *J. Molec. Biol.* **17**, 503.

Cohen, A. S. (1964). *Arthritis Rheum.* **7**, 490.

Cohen, L., Djordjevich, J. and Ormiste, V. (1964). *J. Lab. clin. Med.* **64**, 355.

Conklin, J. L., Dewey, M. M. and May, B. (1962). *J. Histochem. Cytochem.* **10**, 365.

Cory, R. P. and Wold, F. (1965). *Fedn Proc. Fedn Am. Socs exp. Biol.* **24**, 594.

Costello, L. A. and Kaplan, N. O. (1963). *Biochim. biophys. Acta* **73**, 658.

Coutinho, H. B., Padhila, M. C. S., Gomes, J. M. and Alves, J. J. A. (1965). *J. Histochem. Cytochem.* **13**, 339.

Croisille, Y. (1962). *C.r. Acad. Sci., Paris* **254**, 2253.

Cuatrecasas, P. and Segal, S. (1966). *Science, N.Y.* **154**, 533.

Cunningham, V. R. and Field, E. J. (1964). *J. Neurochem.* **11**, 281.

Cunningham, V. R. and Rimer, J. G. (1963). *Biochem. J.* **89**, 50P.

Cunningham, V. R., Phillips, J. and Field, E. J. (1965). *J. clin. Path.* **18**, 765.

Davidson, R. G., Nitowsky, H. M. and Childs, B. (1963). *Proc. natn. Acad. Sci. U.S.A.* **50**, 481.

Davidson, R. G., Fildes, R. A., Glen-Bott, A. M., Harris, H., Robson, E. B. and Cleghorn, T. E. (1965). *Ann. hum. Genet.* **29**, 5.

Davies, D. R. (1934). *Biochem. J.* **28**, 529.

Davies, R. O., Marton, A. V. and Kalow, W. (1960). *Can. J. Biochem.* **38**, 545.

Dawson, D. M., Goodfriend, T. L. and Kaplan, N. O. (1964). *Science, N.Y.* **143**, 929.

Dawson, D. M., Eppenburger, H. M. and Kaplan, N. O. (1965). *Biochem. biophys. Res. Commun.* **21**, 346.

Decker, L. E. and Rau, E. M. (1963). *Proc. Soc. exp. biol. Med.* **112**, 144.

de Grouchy, J. (1958). *Rev. fr. Étud. clin. biol.* **3**, 881.

Delbruck, A., Zebe, E. and Bucher, T. (1959a). *Biochem. Z.* **331**, 273.

Delbruck, A., Schimassek, H., Bartsch, K. and Bucher, T. (1959b). *Biochem. Z.* **331**, 297.

Delcourt, A. and Delcourt, R. (1953). *C.r. Acad. Sci., Paris* **147**, 1104.

Denis, L. J. and Prout, G. R. (1962). Cited in Wieme (1965) p. 338.

Denis, L. J., Prout, G. R. and Woolard, V. (1963). *Invest. urol.* **1**, 101.

Denis, L. J., Prout, G. R., Van Camp, K. and Van Sande, M. (1962). *J. Urol.* **88**, 77.

Deul, D. H. and Van Breeman, J. F. L. (1964). *Clinca chim. Acta* **10**, 276.

de Vaux St. Cyr, C., Hermann, G. and Talal, N. (1963). *Rev. Étud. clin. biol.* **8**, 445.

Dewey, M. M. and Conklin, J. (1960). *Proc. Soc. exp. biol. N.Y.* **105**, 492.

Dioguardi, N. and Agostini, A. (1965). *Enzymol. biol. clin.* **5**, 3.

Dioguardi, N., Agostini, A. and Fiorelli, G. (1962). *Enzymol. biol. clin.* **2**, 116.

Dioguardi, N., Agostini, A., Fiorelli, G. and Lomanto, B. (1963). *J. Lab. clin. Med.* **61**, 713.

Dioguardi, N., Agostini, A., Fiorelli, G. and Mannucci, P. M. (1964). *Enzymol. biol. clin.* **4**, 31.

Di Sabato, G. and Kaplan, N. O. (1964). *J. biol. Chem.* **239**, 438.
Di Sabato, G. and Kaplan, N. O. (1965). *J. biol. Chem.* **240**, 1072.
Di Sabato, G., Pesce, A, and Kaplan, N. O. (1963). *Biochim. biophys. Acta*, **77**, 133.
Downey, W. K. and Andrews, P. (1965). *Biochem. J.* **94**, 642.
Dorn, G. (1965). *Science, N.Y.* **150**, 1183.
Doy, C. H. and Brown, K. D. (1965). *Biochim. biophys. Acta* **104**, 377.
Dreyfus, J. C., Demos, J., Schapira, F. and Schapira, G. (1962). *C.r. Acad. Sci., Paris* **254**, 4384.
Dreiling, D. A., Janowitz, H. D. and Josephberg, L. J. (1963). *Ann. intern. Med.* **58**, 235.
Dubach, U. C. (1961). *Schweiz. med. Wschr.* **92**, 1436.
Dubbs, C. A. (1966). *Clin. Chem.* **12**, 181.
Dubbs, C. A., Vivonia, C. and Hilburn, J. M. (1960). *Science, N.Y.* **131**, 1529.
Dubbs, C. A., Vivonia, C. and Hilburn, J. M. (1961). *Nature, Lond.* **191**, 1203.
Dymling, J. F. (1966). *Scand. J. clin. Lab. Invest.* **18**, 129.
Ecobichan, D. J. (1965). *Can. J. Biochem. Physiol.* **43**, 595.
Ecobichan, D. J. (1966a). *Can. J. Biochem. Physiol.* **44**, 1277.
Ecobichan, D. J. (1966b). *Can. J. Biochem. Physiol.* **44**, 225.
Ecobichan, D. J. and Kalow, W. (1961). *Can. J. Biochem. Physiol.* **39**, 1329.
Ecobichan, D. J. and Kalow, W. (1962). *Biochem. Pharmac.* **11**, 573.
Ecobichan, D. J. and Kalow, W. (1964). *Can. J. Biochem. Physiol.* **42**, 277.
Ecobichan, D. J. and Kalow, W. (1965). *Can. J. Biochem. Physiol.* **43**, 73.
Elliot, B. A. and Fleming, A. F. (1965). *Br. med. J.* **i**, 626.
Elliot, B. A. and Wilkinson, J. H. (1961). *Lancet* **i**, 698.
Elliot, B. A. and Wilkinson, J. H. (1963). *Clin. Sci.* **24**, 343.
Elliot, B. A., Jepson, E. M. and Wilkinson, J. H. (1962). *Clin. Sci.* **23**, 305.
Emerson, P. M., Wilkinson, J. H. and Withycombe, W. A. (1964). *Proc. Ass. clin. Biochem.* **3**, 5.
Emerson, P. M. and Wilkinson, J. H. (1965). *J. clin. Path.* **18** 803.
Emerson P. M., Wilkinson, J. H. and Withycombe, W. A. (1964a). *Nature, Lond.* **202**, 1337.
Emery, A. E. H. (1964). *Nature, Lond.* **201**, 1044.
Emery, A. E. H., Sherbourne, D. H. and Pusch, A. (1965). *Archs Neurol., Chicago* **12**, 251.
Englisova, M., Englis, M., Kovrilek, K. and Masek, K. (1964). *Lancet* **i**, 885.
Eppenburger, H. M., Eppenburger, M., Richterich, R. and Aebi, H. (1964). *Dev. biol.* **10**, 1.
Epstein, C. J., Carter, M. M. and Goldberger, R. F. (1964). *Biochim. biophys. Acta* **92**, 391.
Eränkö, O., Kokko, A. and Söderholm, U. (1962). *Nature, Lond.* **193**, 778.
Eränkö, O., Härkönen, K. and Räisänen, L. (1964). *J. Histochem. Cytochem.* **12**, 570.
Estborn, B. (1959). *Nature, Lond.* **184**, 1636.
Estborn, B. (1961). *Clinica chim. Acta* **6**, 22.
Estborn, B. (1964). *Z. klin. Chem.* **2**, 53.
Estborn, B. and Swedin, B. (1959). *Scand. J. clin. Lab. Invest.* **11**, 235.
Evans, F. T., Gray, P. W. S., Lehmann, H. and Silk, E. (1952). *Lancet* **i**, 1229.
Fan, D. P., Schleisenger, M. J., Torriani, A., Barrett, K. J. and Levinthal, C. (1966). *J. Molec. Biol.* **15**, 32.
Fellenberg, R. V., Eppenburger, H. M., Burger, A., Richterich, R. and Aebi, H. (1963). *Helv. physiol. pharmac. Acta* **21**, C13.
Fieldhouse, B. and Masters, C. J. (1966). *Biochim. biophys. Acta* **118**, 538.
Fildes, R. A. and Harris, H. (1966). *Nature, Lond.* **209**, 261.
Fildes, R. A. and Parr, C. W. (1963). *Nature, Lond.* **201**, 890.

I*

Fincham, J. R. S. (1962). *J. molec. Biol.* **4**, 257.

Fincham, J. R. S. and Coddington, A. (1963). *J. molec. Biol.* **6**, 361.

Fine, I. H. and Costello, L. A. (1963). "Methods in Enzymology," p. 958, Vol. 6. Academic Press, New York.

Fine, I. H., Kaplan, N. O. and Kuftinec, D. (1963). *Biochemistry, N.Y.* **2**, 116.

Fishman, W. H. and Kreischer, J. H. (1963). *Ann. N.Y. Acad. Sci.* **103**, 951.

Fishman, W. H., Green, S. and Inglis, N. I. (1963). *Nature, Lond.* **198**, 685.

Fishman, W. H., Inglis, N. I. and Krant, M. J. (1965). *Clinica. chim. Acta* **12**, 298.

Fleischer, G. A., Potter, C. S. and Wakim, K. G. (1960). *Proc. Soc. exp. Biol. Med.* **103**, 229.

Flexner, L. B., Flexner, J. B., Roberts, R. B. and De La Haba, C. (1960). *Dev. Biol.* **2**, 313.

Fling, M., Horowitz, N. H. and Heinemann, S. F. (1963). *J. biol. Chem.* **238**, 2045.

Flodin, P. (1962). "Dextran gels and their applications in gel filtration". Pharmacia, Uppsala.

Fondy, T. P., Pesce, A., Freeburg, I., Stolzenbach, F. and Kaplan, N. O. (1964). *Biochemistry, N.Y.* **3**, 552.

Fondy, T. P., Everse, J., Driscoll, G. A., Castillo, F., Stolzenbach, F. E. and Kaplan, N. O. (1965). *J. biol. Chem.* **240**, 4219.

Forbat, A., Lehmann, H. and Silk, E. (1953). *Lancet* **ii**, 1067.

Fottrell, P. F. (1966). *Nature, Lond.* **210**, 198.

Freier, E. F. and Bridges, R. A. (1964). *Biochim. biophys. Res. Commun.* **17**, 335.

Fritz, P. J. (1965). *Science, N.Y.* **150**, 364.

Fritz, P. J. and Jacobson, K. B. (1963). *Science, N.Y.* **140**, 64.

Fritz, P. J. and Jacobson, K. B. (1965). *Biochemistry, N.Y.* **4**, 282.

Frydenberg, O. and Nielson, G. (1965). *Hereditas* **54**, 123.

Furth, A. J. and Robinson, D. (1965). *Biochem. J.* **97**, 59.

Futterman, S. and Kinoshita, J. H. (1959). *J. biol. Chem.* **234**, 3174.

Gahne, B. (1963). *Nature, Lond.* **199**, 305.

Gahne, B. (1965). *Genetics, N.Y.* **53**, 681.

Garbus, J., Highman, B. and Altland, P. D. (1964). *Am. J. Physiol.* **207**, 467.

Geiger, H. K. and Mitchell, H. K. (1966). *J. Insect Physiol.* **12**, 747.

Gelderman, A. H. and Peacock, A. C. (1965). *Biochemistry, N.Y.* **4**, 1511.

Gelderman, A. H., Gelboin, H. V. and Peacock, A. C. (1965). *J. Lab. clin. Med.* **65**, 132.

Gelotte, B. (1964). *Acta chem. scand.* **18**, 1283.

Gerhardt, W. and Petri, C. (1965). *Acta neurol. scand.* **41**, Supp. 13, 609.

Gerhardt, W., Clausen, J. and Anderson, H. (1963a). *Acta neurol. scand.* **39**, 31.

Gerhardt, W., Clausen, J., Christensen, E. and Riishede, J. (1963b). *Acta neurol. scand.* **39**, 85.

German, J. L., Evans, V. J., Cortner, J. A. and Westfall, B. B. (1964). *J. natn. Cancer Inst.* **32**, 681.

Giblett, E. R. and Scott, N. M. (1965). *Am. J. hum. Genet.* **17**, 425.

Gibson, D. M., Davisson, E. O., Bachhawat, B. K., Ray, B. R. and Vestling, C. S. (1953). *J. biol. Chem.* **203**, 397.

Gilbert, L. I. and Huddleston, C. J. (1965). *J. Insect Physiol.* **11**, 177.

Gilman, A. and Koelle, G. B. (1949). *Pharmac. Rev.* **1**, 166.

Glassman, E. and Saverance, P. (1963). *J. Elisha Mitchell Scient. Soc.* **79**, 139.

Goebelsman, V. and Beller, F. K. (1965). *Z. klin. Chem.* **3**, 49.

Goldberg, A. F., Takakura, K. and Rosenthal, R. L. (1966). *Nature, Lond.* **211**, 41.

Goldberg, E. (1963). *Science, N.Y.* **139**, 602.

Goldberg, E. (1964). *Ann. N.Y. Acad. Sci.* **121**, 560.

Goldberg, E. (1965b). *Arch Biochem. Biophys.* **109**, 134.

Goldberg, E. (1965a). *Science, N.Y.* **148**, 391.

Goldberg, E. (1966). *Science, N.Y.* **151**, 1091.

Goldberg, E. and Cather, J. N. (1963). *J. cell comp. Physiol.* **41**, 31.

Goldman, R. D., Kaplan, N. O. and Hall, T. C. (1964). *Cancer Res.* **24**, 389.

Gomori, G. (1950). *Stain Technol.* **25**, 81.

Gonzalez, C., Ureta, T., Sanchez, R. and Niemeyer, H. (1964). *Biochem. biophys. Res. Commun.* **16**, 347.

Goodfriend, T. and Kaplan, N. O. (1963). *J. cell Biol.* **19**, 28A.

Goodfriend, T. L. and Kaplan, N. O. (1964). *J. biol. Chem.* **239**, 130.

Goodfriend, T. L., Sokol, D. M. and Kaplan, N. O. (1966). *J. molec. Biol.* **15**, 18.

Gordon, A. H., Keil, B. and Sebasta, K. (1949). *Nature, Lond.* **164**, 498.

Gotts, R. and Skendzel, L. P. (1966). *Clinica chim. Acta* **14**, 505.

Grabar, P. (1958). "Advances in Protein Chemistry," p. 1, Vol. 13. Academic Press, New York.

Grabar, P. and Williams, C. A. (1953). *Biochim. biophys. Acta* **10**, 193.

Graymore, C. N. (1964). *Nature, Lond.* **201**, 615.

Graymore, C. N. (1965). *In* "First Intern. Symp. Biochemistry of the Retina." London, 1964, p. 83. (C. N. Graymore, ed.). Academic Press, London.

Greiling, H., Engels, G. and Kisters, R. (1964). *Klin. Wschr.* **42**, 427.

Grell, E. H., Jacobson, K. B. and Murphy, J. B. (1965). *Science, N.Y.* **149**, 80.

Grimes, H. and Fottrell, P. F. (1966). *Nature, Lond.* **212**, 294.

Grimm, F. C. and Doherty, D. G. (1961). *J. biol. Chem.* **236**, 1980.

Grossbard, L. and Schimke, R. T. (1966). *J. biol. Chem.* **241**, 3546.

Grossberg, A. L., Harris, E. H. and Schlamowitz, M. (1961). *Archs Biochem.* **93**, 267.

Grunder, A. A., Sartore, G. and Stormont, C. (1965). *Genetics, Princeton* **52**, 1345.

Grundig, E., Czitober, H. and Schobel, B. (1965). *Clinica chim. Acta* **12**, 157.

Gutman, A. B. and Jones, B. (1949). *Proc. Soc. exp. Biol. Med.* **71**, 572.

Güttler, F. and Clausen, J. (1965). *Enzymol. biol. clin.* **5**, 55.

Haije, W. G. and de Jong, M. (1963). *Clinica chim. Acta* **8**, 620.

Hansson, A., Johannson, B. and Sievers, J. (1962). *Lancet,* **i**, 167.

Hardy, S. M. (1965). *Nature, Lond.* **206**, 933.

Harris, H. and Robson, E. B. (1963). *Biochim. biophys. Acta* **73**, 649.

Harris, H. and Whittaker, M. (1959). *Nature, Lond.* **183**, 1808.

Harris, H. and Whittaker, M. (1961). *Nature, Lond.* **191**, 496.

Harris, H. and Whittaker, M. (1962a). *Ann. hum. Genet.* **26**, 73.

Harris, H. and Whittaker, M. (1962b). *Ann. hum. Genet.* **26**, 59.

Harris, H., Hopkinson, D. A. and Robson, E. B. (1962). *Nature, Lond.* **196**, 1296.

Harris, H., Hopkinson, D. A., Robson, E. B. and Whittaker, M. (1963). *Ann. hum. Genet.* **26**, 359.

Harris, H., Whittaker, M., Lehmann, H. and Silk, E. (1960). *Acta genet. statist. med.* **10**, 1.

Haupt, I. and Giersberg, H. (1958). *Naturwissenschaften* **45**, 268.

Hawkins, D. F. and Whyley, G. A. (1966). *Clin. chim. Acta* **13**, 713.

Hawrylewicz, E. J. and Blair, W. H. (1965). *Aerospace Med.* **36**, 369.

Hayman, S. and Alberty, R. A. (1961). *Ann. N.Y. Acad. Sci.* **94**, 812.

Henderson, N. S. (1965). *J. exp. Zool.* **158**, 263.

Henderson, N. S. (1966). *Archs Biochem. Biophys.* **117**, 28.

Henson, C. P. and Cleland, W. W. (1964). *Biochemistry, N.Y.* **3**, 338.

Herbert, F. K. (1944). *Biochem. J.* **38**, xxiii.
Herbert, F. K. (1945). *Biochem. J.* **39**, iv.
Herbert, F. K. (1946). *Q. Jl. Med.* **15**, 221.
Hermann, G., Talal, N., De Vaux St. Cyr, Ch. and Escribano, J. (1963). "Protides of the Biological Fluids". Xth Colloquim, Bruges, p. 186. Elsevier, Amsterdam.
Hermans, P. E., McGuckin, W. F., McKenzie, B. F. and Bayrd, E. D. (1960). *Proc. Staff meet. Mayo Clin.* **35**, 792.
Hess, B. (1958). *Ann. N.Y. Acad. Sci.* **75**, 292.
Hess, B. and Walter, S. I. (1960). *Klin. Wschr.* **38**, 1080.
Hess, B. and Walter, S. I. (1961). *Ann. N.Y. Acad. Sci.* **94**, 890.
Hess, A. R., Angel, R. W., Barron, K. D. and Bernsohn, J. (1963). *Clin. chim. Acta* **8**, 656.
Hill, B. R. (1958). *Ann. N.Y. Acad. Sci.* **75**, 304.
Hill, B. R. and Levi, C. (1954). *Cancer Res.* **14**, 513.
Hinks, M. and Masters, C. J. (1964). *Biochemistry, N.Y.* **3**, 1789.
Hirs, C. H. W., Stein, W. H. and Moore, S. (1951). *J. Am. Chem. Soc.* **73**, 1893.
Hirschfeld, J. (1960). *Nature, Lond.* **186**, 321.
Hirschman, S. Z., Morell, A. G. and Scheinberg, I. H. (1961). *Ann. N.Y. Acad. Sci.* **94**, 960.
Hochachka, P. W. (1965). *Archs Biochem. Biophys.* **111**, 96.
Hochachka, P. W. (1966). *Comp. Biochem. Physiol.* **18**, 261.
Hodson, A. W., Latner, A. L. and Masterton, J. (1965). *Archs oral Biol.* **10**, 547.
Hodson, A. W., Latner, A. L. and Raine, L. (1962). *Clin. chim. Acta* **7**, 255.
Hodson, A. W., Latner, A. L., Raine, L. and Skillen, A. W. (1961). *J. Physiol. Lond.* **159**, 54P.
Holbrook, J. J. and Pfleiderer, G. (1965). *Biochem. Z.* **342**, 111.
Hopkinson, D. A. and Harris, H. (1965). *Nature, Lond.* **208**, 410.
Hopkinson, D. A., Spencer, N. and Harris, H. (1963). *Nature, Lond.* **199**, 969.
Hopkinson, D. A., Spencer, N. and Harris, H. (1964). *Am. J. hum. Genet.* **16**, 151.
Hopsu, V. K. and Glenner, G. G. (1964). *J. Histochem. Cytochem.* **12**, 674.
Hopsu, V. K., Santhi, R. S. and Glenner, G. G. (1965). *J. Histochem. Cytochem.* **13**, 117.
Hori, J. (1966). *Analyt. Biochem.* **16**, 234.
Hsieh, K. M., Suntzeff, V. and Cowdry, E. V. (1955). *Proc. Soc. exp. Biol. Med.* **89**, 627.
Hudock, G. A., Mellin, D. B. and Fuller, R. C. (1965). *Science, N.Y.* **150**, 776.
Hule, V. (1966). *Clin. chim. Acta* **13**, 431.
Hunter, R. L. and Burstone, M. S. (1960). *J. Histochem. Cytochem.* **8**, 58.
Hunter, R. L. and Markert, C. L. (1957). *Science, N.Y.* **125**, 1294.
Hunter, R. L. and Strachan, D. S. (1961). *Ann. N.Y. Acad. Sci.* **94**, 861.
Hunter, R. L., Denuce, J. M. and Strachan, D. S. (1961). *Annls Histochem.* **6**, 447.
Hunter, R. L., Rocha, J. T., Pfrender, A. R. and De Jong, D. C. (1964). *Ann. N.Y. Acad. Sci.* **121**, 532.
Jacobs, H., Heldt, H. W. and Klingenberg, M. (1964). *Biochem. biophys. Res. Commun.* **16**, 516.
Jaenicke, R. (1963). *Biochem. Z.* **338**, 614.
Jaenicke, R. (1964). *Biochim. biophys. Acta* **85**, 186.
Jaenicke, R. and Pfleiderer, G. (1962). *Biochim. biophys. Acta* **60**, 615.
Jensen, K. and Thorling, E. B. (1965). *Acta path. microbiol. scand.* **63**, 385.
Johnson, F. M. (1966). *Nature, Lond.* **212**, 843.

Johnson, F. M. and Denniston, C. (1964). *Nature, Lond.* **204**, 906.

Johnson, F. M. and Sakai, R. (1964). *Nature, Lond.* **203**, 373.

Johnson, H. L. and Kampschmidt, R. F. (1965). *Proc. Soc. exp. Biol. Med.* **120**, 557.

Johnston, H. A., Wilkinson, J. H., Withycombe, W. A. and Raymond, S. (1965). *J. clin. Path.* **19**, 250.

Jolley, R. L. and Mason, H. S. (1965). *J. biol. Chem.* **240**, PC1489.

Joseph, R. R., Olivero, M. S. and Ressler, N. (1966). *Gastroenterology* **51**, 377.

Kaji, A., Trayser, K. A. and Colowick, S. P. (1961). *Ann. N.Y. Acad. Sci.* **94**, 798.

Kalbag, R. M., Park, D. C. and Pennington, R. J. (1966). *Proc. ass. clin. Biochem.* **IV**, 88.

Kalow, W. (1959). "Biochemistry of Human Genetics", p. 39. Churchill, London.

Kalow, W. and Davies, R. O. (1958). *Biochem. Pharmac.* **1**, 183.

Kalow, W. and Genest, K. (1957). *Can. J. Biochem. Physiol.* **35**, 339.

Kalow, W. and Gunn, D. R. (1958). *Ann. hum. Genet.* **23**, 239.

Kalow, W. and Staron, N. (1957). *Can. J. Biochem. Physiol.* **35**, 1305.

Kaminski, M. (1966). *Nature, Lond.* **209**, 723.

Kaminski, M. and Gajos, E. (1964). *Nature, Lond.* **201**, 716.

Kanungo, M. S. and Singh, S. N. (1965). *Biochem. biophys. Res. Commun.* **21**, 454.

Kaplan, N. O. (1961). *In* "Mechanism of Action of Steroid Hormones," p. 247. (C. A. Villee and A. A. Engel, eds.). Pergamon Press, Oxford.

Kaplan, N. O. (1963). *Bact. Rev.* **27**, 155.

Kaplan, N. O. (1964). *Brookhaven Symp. Biol.* No. 17, 131.

Kaplan, N. O. (1965). *In* "Evolving Genes and Proteins," p. 243. (V. Bryson and H. J. Vogel, eds.). Academic Press, New York.

Kaplan, N. O. and Cahn, R. D. (1962). *Proc. natn. Acad. Sci. U.S.A.* **48**, 2123.

Kaplan, N. O. and Ciotti, M. M. (1961). *Ann. N.Y. Acad. Sci.* **94**, 701.

Kaplan, N. O. and White, S. (1963). *Ann. N.Y. Acad. Sci.* **103**, 835.

Kaplan, N. O., Ciotti, M. M., Hamolsky, M. and Bieber, R. E. (1960). *Science, N.Y.* **131**, 392.

Kar, N. C. and Pearson, C. M. (1963). *Proc. natn. Acad. Sci. U.S.A.* **50**, 995.

Kar, N. C. and Pearson, C. M. (1964). *Proc. Soc. exp. Biol. Med.* **116**, 733.

Kar, N. C. and Pearson, C. M. (1965). *Am. J. clin. Path.* **43**, 207.

Kattamis, C., Zannos-Mariolea, L., Franco, A. P., Liddell, J., Lehmann, H. and Davies, D. (1962). *Nature, Lond.* **196**, 599.

Katunama, N., Tomino, I. and Nishino, H. (1966). *Biochem. biophys. Res. Commun.* **22**, 321.

Katz, A. M. and Kalow, W. (1965). *Can. J. Biochem. Physiol.* **43**, 1653.

Katz, A. M. and Kalow, W. (1966). *Nature, Lond.* **209**, 1350.

Katzen, H. M. (1966). *Biochem. biophys. Res. Commun.* **24**, 531.

Katzen, H. M. and Schimke, R. T. (1965). *Proc. natn. Acad. Sci. U.S.A.* **54**, 1218.

Katzen, H. M., Soderman, D. D. and Nitowsky, H. M. (1965). *Biochem. biophys. Res. Commun.* **19**, 377.

Kaufman, L., Lehmann, H. and Silk, E. (1960). *Br. Med. J.* **i**, 166.

Kawashima, N., Hyodo, H. and Uritani, I. (1964). *Phytopathology* **54**, 1086.

Kazazaian, H. H., Young, W. J. and Childs, B. (1965). *Science, N.Y.* **150**, 1601.

Keck, K. (1961). *Ann. N.Y. Aca. Sci.* **94**, 741.

Keck, K. and Choules, E. A. (1962). *Arch. biochem. Biophys.* **99**, 205.

Keck, K. and Choules, E. A. (1964). *J. cell Biol.* **18**, 459.

Keiding, N. R. (1959). *Scand. J. clin. Lab. Invest.* **11**, 106.

Keiding, N. R. (1964). *Clin. Sci.* **26**, 292.

Keiding, N. R. (1966). *Scand. J. clin. Lab. Invest.* **18**, 134.
Keller, E. C., Saverance, P. and Glassman, E. (1963). *Nature, Lond.* **198**, 286.
Kikkawa, H. (1960). *Rep. Scient. Wks.* Osaka Univ. **2**, 41.
Kikkawa, H. and Ogita, Z. (1962). *Jap. J. Genet.* **37**, 394.
Kim, H. C., D'Ioro, A. D. and Paik, W. K. (1966). *Can. J. Biochem. Physiol.* **44**, 103.
King, C. M. and Gutmann, H. R. (1964). *Cancer Res.* **24**, 770.
King, E. J., Wood, E. J. and Delory, G. E. (1945). *Biochem. J.* **39**, xxiv.
Kirkman, H. N. (1962). *Am. J. Dis. Child.* **104**, 566.
Kirkman, H. N. and Hendrickson, E. M. (1962). *J. biol. Chem.* **237**, 2371.
Kirkman, H. N. and Hendrickson, E. M. (1963). *Am. J. hum. Genet.* **15**, 241.
Kirkman, H. N. and Riley, H. D. (1961). *Am. J. Dis. Child.* **102**, 313.
Kirkman, H. N., Rosenthal, I. M., Simon, E. R. and Carson, P. E. (1963). *J. Paed.* **63**, 719.
Kirkman, H. N., Rosenthal, E. R., Simon, E. R., Carson, P. E. and Brinson, A. G. (1964a). *J. Lab. clin. Med.* **63**, 715.
Kirkman, H. N., Schettini, F. and Pickard, B. M. (1964b). *J. Lab. clin. Med.* **63**, 726.
Kirkman, H. N., McCurdy, P. R. and Naiman, J. L. (1964c). *Cold Spring Harb. Symp. Quant. Biol.* **29**, 391.
Kirkman, H. N., Simon, E. R. and Pickard, B. M. (1965). *J. Lab. clin. Med.* **66**, 834.
Kitchener, P. N., Neale, F. C., Posen, S., Brudenell-Woods, J. (1965). *Am. J. clin. Path.* **44**, 654.
Klapper, M. H. and Hackett, D. P. (1965). *Biochim. biophys. Acta* **96**, 272.
Koen, A. L. and Shaw, C. R. (1964). *Biochem. biophys. Res. Commun.* **15**, 92.
Koen, A. L. and Shaw, C. R. (1966). *Biochim. biophys. Acta* **90**, 231.
Kohn, J. (1957a). *Biochem. J.* **65**, 9P.
Kohn, J. (1957b). *Clinica chim. Acta* **2**, 297.
Kohn, J. (1960). "Chromatographic and Electrophoretic Techniques", Vol. II, p. 56. (I. Smith, ed.). Wm. Heinemann, London.
Koler, R. D., Bigley, R. H., Jones, R. T., Rigas, D. A., Van Kellinghen, P. and Thompson, P. (1964). *Cold Spring Harb. Symp. Quant. Biol.* **29**, 213.
Komatsu, S. and Michaelis, M. (1966). *Proc. Soc. exp. Biol. Med.* **121**, 1028.
Komma, D. J. (1963). *J. Histochem. Cytochem.* **11**, 619.
Kontinnen, A. (1961). *Lancet* **ii**, 556.
Kontinnen, A. and Halonen, P. I. (1962). *Am. J. Cardiol.* **10**, 525.
Kontinnen, A. and Lindy, S. (1965). *Nature, Lond.* **208**, 782.
Korhonen, L. K. and Korhonen, E. (1965). *Histochemie.* **5**, 279.
Kormendy, K., Gantner, G. and Hamm, R. (1965). *Biochem. Z.* **342**, 31.
Korner, N. H. (1962). *J. clin. Path.* **15**, 195.
Kotzaurek, R. and Schobel, B. (1963). *Klin. Wschr.* **41**, 956.
Kowlessar, O. D., Haeffner, L. J. and Riley, E. M. (1961). *Ann. N.Y. Acad. Sci.* **94**, 836.
Kowlessar, O. D., Haeffner, L. J. and Sleisinger, M. H. (1960). *J. clin. Invest.* **39**, 671.
Kowlessar, O. D., Pert, J. H., Haeffner, L. J. and Sleisinger, M. H. (1958). *Proc. Soc. exp. Biol. Med.* **100**, 191.
Kraus, A. P. and Neely, C. L. (1964). *Science, N.Y.* **145**, 595.
Kreischer, J. H., Close, V. A. and Fishman, W. H. (1965). *Clinica chim. Acta* **11**, 122.
Kreutzer, H. H. and Eggels, P. H. (1965). *Clinica chim. Acta* **12**, 80.
Kreutzer, H. H. and Fennis, W. H. S. (1964). *Clinica chim. Acta* **9**, 64.
Kreutzer, H. H. and Jacobs, Ph. (1965). *Clinica chim. Acta* **11**, 184.
Kreutzer, H. H., Jacobs, Ph. and Francke, C. (1965). *Clinica chim. Acta* **11**, 159.

Kroner, H. and Reunhaver, R. (1964). *Z. Physiol. Chemie.* **336**, 227.

Kun, E. and Volfin, P. (1966). *Biochem. biophys. Res. Commun.* **22**, 187.

Kusakabe, T. and Miyake, T. (1966). *J. clin. Endocr. Metab.* **26**, 615.

Labremont, E. N. and Schrader, R. M. (1964). *Nature, Lond.* **204**, 883.

Lai, L., Nevo, S. and Steinberg, A. G. (1964). *Science, N.Y.* **145**, 1187.

La Motta, R. V., McComb, R. B. and Wetstone, H. J. (1965). *Can. J. Physiol. Pharmac.* **43**, 313.

Langman, M. J. S., Lenthold, E., Robson, E. B., Harris, J., Luffman, J. E. and Harris, H. (1966). *Nature, Lond.* **212**, 41.

Latner, A. L. (1962). *Proc. Ass. clin. Biochem.* **2**, 6.

Latner, A. L. (1963). *Clin. Chem.* **9**, 478.

Latner, A. L. (1964). *Proc. Ass. clin. Biochem.* **3**, 120.

Latner, A. L. (1965). "Enzymes in Clinical Chemistry". p. 110. (R. Ruyssen and L. Vandendriessche, eds.). Elsevier Publishing Company, Amsterdam.

Latner, A. L. (1966). "The Binding of Circulating Enzymes by Plasma Proteins." Proceedings Vth West European Symposium on Clinical Chemistry, Paris.

Latner, A. L. and Skillen, A. W. (1961b). *Proc. Ass. clin. Biochem.* **1**, 129.

Latner, A. L. and Skillen, A. W. (1961a). *Lancet* **ii**, 1286.

Latner, A. L. and Skillen, A. W. (1962). *Proc. Ass. clin. Biochem.* **2**, 3.

Latner, A. L. and Skillen, A. W. (1963). *Proc. Ass. clin. Biochem.* **2**, 100.

Latner, A. L. and Skillen, A. W. (1964). *J. Embryol. exp. Morph.* **12**, 501.

Latner, A. L. and Turner, D. M. (1963). *Lancet* **i**, 1293.

Latner, A. L. and Turner, D. M. (1967). *Clin. chim. Acta* **15**, 97.

Latner, A. L., Siddiqui, S. A. and Skillen, A. W. (1966a). *Science, N.Y.* **154**, 527.

Latner, A. L., Skillen, A. W. and Swinney, J. (1966b). Proceedings European Dialysis and Transplant Assoc. Vol. II., p. 187. Excerpta Medica Foundation.

Latner, A. L., Turner, D. M. and Way, S. A. (1966c). *Lancet* **ii**, 814.

Latner, A. L., Gardner, P. S., Turner, D. M. and Brown, J. O. (1965). *Lancet* **i**, 197.

Laufer, H. (1960). *Ann. N.Y. Acad. Sci.* **89**, 490.

Laufer, H. (1961). *Ann. N.Y. Acad. Sci.* **94**, 825.

Laufer, H. (1963). *Ann. N.Y. Acad. Sci.* **103**, 1137.

Laurent, G., Charrel, M., Castay, M., Nahan, D., Marriq, C. and Derrien, Y. (1962). *C.r. Séance. Soc. Biol.* **156**, 1461.

Lauryssens, M. G., Lauryssens, M. J. and Zondag, H. A. (1964). *Clin. chim. Acta* **9**, 276.

Law, G. R. J. and Munro, S. S. (1965). *Science, Lond.* **149**, 1518.

Lawrence, S. H. (1964). "The Zymogram in Clinical Medicine". Charles C. Thomas Springfield, Illinois.

Lawrence, S. H. and Melnick, P. J. (1961). *Proc. Soc. exp. Biol. Med.* **107**, 998.

Lawrence, S. H., Melnick, P. J. and Weimer, H. E. (1960). *Proc. Soc. exp. Biol. Med.* **105**, 572.

Leeper, R. A. (1963). *J. clin. Endocr.* **23**, 426.

Lehmann, H. and Liddell, J. (1964). *In* "Progress in Medical Genetics". Vol. III, p. 75. (A. G. Steinberg and A. G. Bearn, eds.). Grune and Stratton, New York and London.

Lehmann, H. and Ryan, E. (1956). *Lancet* **ii**, 124.

Lehmann, H. and Silk, E. (1953). *Br. med. J.* **1**, 767.

Lehmann, H., Paston, V. and Ryan, E. (1958). *J. clin. Path.* **11**, 554.

Lehmann, H., Liddell, J., Blackwell, B., O'Connor, D. C. and Daws, A. V. (1963). *Br. med. J.* **i**, 1116.

Levinthal, C., Signer, E. R. and Fetherolf, K. (1962). *Proc. natn. Acad. Sci. U.S.A.* **48**, 1230.

Lewis, A. A. M. and Hunter, R. L. (1966). *J. Histochem. Cytochem.* **14**, 33.

Liddell, J., Lehmann, H., Davies, D. and Sharih, A. (1962a). *Acta genet. Statist. med.* **13**, 95.

Liddell, J., Lehmann, H., Davies, D. and Sharih, A. (1962b). *Lancet* **i**, 463.

Linder, D. and Gartler, S. M. (1965a). *Science, N.Y.* **150**, 67.

Linder, D. and Gartler, S. M. (1965b). *Am. J. hum. Genet.* **17**, 212.

Lindsay, D. T. (1963). *J. exp. Zool.* **152**, 75.

Lindskog, S. (1960). *Biochim. biophys. Acta* **39**, 218.

Lindy, S. and Kontinnen, A. (1966a). *Nature, Lond.* **209**, 79.

Lindy, S. and Kontinnen, A. (1966b). *Clin. chim. Acta* **14**, 615.

Lindy, S. and Rajasalmi, M. (1966). *Science, N.Y.* **153**, 1401.

Long, W. K., Kirkman, H. N. and Sutton, H. E. (1965). *J. Lab. clin. Med.* **65**, 81.

Lowenstein, J. M. and Smith, S. R. (1962). *Biochim. biophys. Acta* **56**, 385.

Lowenthal, A., Van Sande, M. and Karcher, D. (1961a). *Pure appl. Chem.* **3**, 411.

Lowenthal, A., Van Sande, M. and Karcher, D. (1961b). *Ann. N.Y. Acad. Sci.* **94**, 988.

Lowenthal, A., Van Sande, M. and Karcher, D. (1961c). *J. Neurochem.* **7**, 135.

Lowenthal, A., Karcher, D. and Van Sande, M. (1964). *J. Neurochem.* **11**, 247.

Luang Eng, L-I. (1966). *Nature, Lond.* **210**, 1183.

Lundin, L-G. and Allison, A. C. (1966). *Biochim. biophys. Acta* **127**, 527.

McCune, D. C. (1961). *Ann. N.Y. Acad. Sci.* **94**, 723.

McGeachin, R. L. and Lewis, J. P. (1959). *J. biol. Chem.* **234**, 795.

McGeachin, R. L. and Potter, B. A. (1961). *Nature, Lond.* **189**, 751.

McGeachin, R. L. and Reynolds, J. M. (1961). *Ann. N.Y. Acad. Sci.* **94**, 996.

MacIntyre, R. J. and Wright, T. R. F. (1966). *Genetics, Princeton* **53**, 371.

MacIntyre, R. (1966). *Genetics, Princeton* **53**, 46.

McKinley-McKee, J. S. and Moss, D. W. (1965). *Biochem. J.* **96**, 583.

McLean, P., Brown, J., Greenslade, K. and Brew, K. (1966). *Biochem. biophys. Res. Commun.* **23**, 117.

McMaster, Y., Tennant, R., Clubb, J. S., Neale, F. C. and Posen, S. (1964). *J. Obstet. Gynaec. Br. Commonw.* **71**, 735.

Macalalag, E. V. and Prout, G. R. (1964). *J. Urol.* **92**, 416.

Mager, M., Blatt, W. F. and Abelmann, W. H. (1966). *Clinica chim. Acta* **14**, 689.

Maisel, H., Keinoger, M. and Syner, F. (1965). *Invest. Opthalmol.* **4**, 362.

Mannucci, P. M. and Dioguardi, N. (1966). *Clinica chim. Acta* **14**, 215.

Manwell, C. (1966). *Comp. Biochem. Physiol.* **17**, 805.

Manwell, C. and Kerst, K. V. (1966). *Comp. Biochem. Physiol.* **17**, 741.

Markert, C. L. (1962). *In* "Hereditary, Developmental and Immunological Aspects of Kidney Disease", Vol. III, p. 54. (J. Metcoff, ed.). Northwestern University Press, Evanston.

Markert, C. L. (1963a). *Science, N.Y.* **140**, 1329.

Markert, C. L. (1963b). *In* "21st Symposium of the Society for Study of Development and Growth", p. 65. (D. Rudnick, ed.). Academic Press, New York.

Markert, C. L., *Int. Cong. Biochem.* **6**(**4**), 320.

Markert, C. L. and Appella, E. (1961). *Ann. N.Y. Acad. Sci.* **94**, 678.

Markert, C. L. and Appella, E. (1963). *Ann. N.Y. Acad. Sci.* **103**, 915.

Markert, C. L. and Faulhaber, I. (1965). *J. exp. Zool.* **159**, 319.

Markert, C. L. and Hunter, R. L. (1959). *J. Histochem. Cytochem.* **7**, 42.

Markert, C. L. and Moller, F. (1959). *Ann. N.Y. Acad. Sci.* **45**, 753.

Markert, C. L. and Ursprung, H. (1962). *Dev. Biol.* **5**, 363.

Marks, P. A., Szeinberg, A. and Banks, J. (1961). *J. biol. Chem.* **236**, 10.

Massaro, E. J. and Markert, C. L. (1966). *Fedn Proc. Fedn Am. Socs. exp. Biol.* **25**, 711.

Masters, C. J. (1963). *Aust. J. Biol. Sci.* **16**, 709.

Masters, C. J. (1964). *Biochim. biophys. Acta* **89**, 161.

Masters, C. J. and Hinks, M. (1966). *Biochim. biophys. Acta* **113**, 611.

Masurovsky, E. B. and Noback, C. R. (1963). *Nature, Lond.* **200**, 847.

Maynard, E. (1964). *J. exp. Zool.* **157**, 251.

Meade, B. W. and Rosalki, S. B. (1962). *Lancet* **i**, 1407.

Meade, B. W. and Rosalki, S. B. (1963). *J. Obstet. Gynec. Br. Commonw.* **70**, 862.

Meade, B. W. and Rosalki, S. B. (1964). *J. clin. Path.* **17**, 61.

Meier, H., Jordan, E. and Hoag, W. G. (1962). *J. Histochem. Cytochem.* **10**, 103.

Meister, A. (1950). *J. natn. Cancer Inst.* **10**, 1263.

Menzel, D. B., Craig, R. and Hoskins, W. M. (1963). *J. Insect Physiol.* **9**, 479.

Meyer, J. A., Garber, E. D. and Shaeffer, S. G. (1964). *Bot. Gaz.* **125**, 298.

Micheli, A. and Grabar, P. (1961). *Annls Inst. Pasteur, Paris* **100**, 569.

Miller, Z. B., Naor, E., Milkovitch, L. and Schmidt, W. M. (1964). *Obstet. Gynec. N.Y.* **24**, 707.

Mitchell, H. K. (1966). *J. Insect Physiol.* **12**, 755.

Mitchell, H. K. and Weber, U. M. (1965). *Science, N.Y.* **148**, 964.

Monis, B. (1964). *J. Histochem. Cytochem.* **12**, 869.

Monis, B. (1965). *Fedn Proc. Fedn Am. Socs. exp. Biol.* **24**, 681.

Moog, F., Vire, H. E. and Grey, R. D. (1966). *Biochim. biophys. Acta* **113**, 366.

Moore, B. W. and Angeletti, P. U. (1961). *Ann. N.Y. Acad. Sci.* **94**, 659.

Moore, B. W. and Lee, R. H. (1960). *J. biol. Chem.* **235**, 1359.

Moore, B. W. and Wortman, B. (1959). *Biochim. biophys. Acta* **34**, 260.

Moore, R. O. and Villee, C. A. (1963). *Science, Lond.* **142**, 389.

Morino, Y., Itoh, H. and Wada, H. (1963). *Biochem. biophys. Res. Commun.* **13**, 348.

Morino, Y., Kagamiyama, H. and Wada, H. (1964). *J. biol. Chem.* **239**, PC943.

Moss, D. W. (1963). *Nature, Lond.* **200**, 1206.

Moss, D. W. (1964). *Proc. Ass. clin. Biochem.* **3**, 132.

Moss, D. W. and King, E. J. (1962). *Biochem. J.* **84**, 192.

Moss, D. W., Campbell, D., Anagnostou-Kakaras, E. and King, E. J. (1961a). *Biochem. J.* **81**, 441.

Moss, D. W., Campbell, D., Anagnostou-Kakaras, E. and King, E. J. (1961b). *Pure appl. Chem.* **3**, 397.

Moustafa, E. (1963). *Nature, Lond.* **199**, 1189.

Moustafa, E. (1964). *N.Z. Jl. Sci.* **7**, 608.

Mull, J. D. and Starkweather, W. H. (1965). *Am. J. clin. Path.* **44**, 231.

Nace, G. W. (1963). *Ann. N.Y. Acad. Sci.* **103**, 980.

Nace, G. W., Suyama, T. and Smith, N. (1961). *In* "Germ cells and earliest stages of development", p. 564. (S. Ranzi, ed.). Baselli, Pavia, Italy.

Nachlas, M. M., Friedman, M. M. and Cohen, S. P. (1964). *Surgery* **55**, 700.

Nakagawa, S. and Tsuji, H. (1966). *Clinica chim. Acta* **13**, 155.

Nakano, E. and Whiteley, A. H. (1965). *J. exp. Zool.* **159**, 167.

Nance, W. E. and Uchida, I. (1964). *Am. J. hum. Genet.* **16**, 380.

Nance, W. E., Claflin, A. and Smithies, O. (1963). *Science, N.Y.* **142**, 1075.

Neale, F. C., Clubb, J. S., Hotchkis, D. and Posen, S. (1965). *J. clin. Path.* **18**, 359.

Nebel, E. J. and Conklin, J. L. (1964). *Proc. Soc. exp. Biol. Med.* **115**, 532.

Neel, J. V., Salzano, F. M., Junqueria, P. C., Keiter, F. and Maybury-Lewis, D. (1964). *Am. J. hum. Genet.* **16**, 52.

Neelin, J. M. (1963). *Can. J. Biochem. Physiol.* **41**, 369.

Nelson, B. D. (1966). *Proc. Soc. exp. Biol. Med.* **121**, 998.
Newton, M. A. (1966). *J. clin. Path.* **19**, 491.
Nielands, J. B. (1952). *J. biol. Chem.* **199**, 373.
Nishimura, E. T., Carson, N. and Kobara, T. Y. (1964). *Archs Biochem. Biophys.* **108**, 452.
Nisselbaum, J. S. and Bodansky, O. (1959). *J. biol. Chem.* **234**, 3276.
Nisselbaum, J. S. and Bodansky, O. (1961b). *J. biol. Chem.* **236**, 323.
Nisselbaum, J. S. and Bodansky, O. (1963). *Ann. N.Y. Acad. Sci.* **103**, 930.
Nisselbaum, J. S. and Bodansky, O. (1964). *J. biol. Chem.* **239**, 4232.
Nisselbaum, J. S. and Bodansky, O. (1965). *Science, Lond.* **149**, 195.
Nisselbaum, J. S. and Bodansky, O. (1966). *J. biol. Chem.* **241**, 2661.
Nisselbaum, J. S., Packer, D. E. and Bodansky, O. (1964) *J. biol. Chem.* **239**, 2830.
Nitowsky, H. M. and Soderman, D. D. (1964). *Expl Cell Res.* **33**, 562.
Nitowsky, H. M., Davidson, R. G., Soderman, D. D. and Childs, B. (1965). *Bull. Johns Hopkins Hosp.* **117**, 363.
Noerby, S. (1965). *Expl Cell Res.* **36**, 663.
Nordentoft-Jensen, B. (1964). *Clin. Sci.* **26**, 299.
Nutter, D. D., Trujillo, N. P. and Evans, J. M. (1966). *Am. Heart J.* **72**, 315.
Nyman, P. O. (1961). *Biochim. biophys. Acta* **52**, 1.
Oger, A. and Bischops, L. (1966). *Clinica chim. Acta* **13**, 670.
Ohno, S., Poole, J. and Gustavsson, I. (1965). *Science, N.Y.* **150**, 1737.
Oki, Y., Oliver, W. T. and Funnell, H. S. (1964). *Nature, Lond.* **203**, 605.
Oort, J. and Willighagen, R. G. J. (1961). *Nature, Lond.* **190**, 642.
Oppenoorth, F. J. and Van Asperen, K. (1960). *Science, N.Y.* **132**, 298.
Orelli, A. V. and Dubach, U. C. (1964). *Klin. Wschr.* **42**, 58.
Ornstein, L. and Davis, B. J. (1961). "Disc electrophoresis". Preprinted by Distillation Products Industries, Eastman Kodak Co., Rochester, New York.
Pagliaro, L. and Notarbartolo, A. (1962). *Lancet* **ii**, 1261.
Pagliaro, L. and Notarbartolo, A. (1962). *Lancet* **i**, 1043.
Papenburg, J., Von Wartburg, J. P. and Aebi, H. (1965). *Biochem. Z.* **342**, 95.
Parr, C. W. (1966). *Nature, Lond.* **210**, 487.
Parr, C. W. and Fitch, L. I. (1964). *Biochem. J.* **93**, 28c.
Parr, C. W. and Parr, I. B. (1965). *Biochem. J.* **95**, 16p.
Patil, S., Evans, H. J. and McMahill, P. (1963). *Nature, Lond.* **200**, 1322.
Patterson, E. K., Hsiao, H. S. and Keppel, A. (1963). *J. biol. Chem.* **238**, 3611.
Patterson, E. K., Hsiao, H. S., Keppel, A. and Sorof, S. (1965). *J. biol. Chem.* **240**, 710.
Paul, J. and Fottrell, P. F. (1961). *Ann. N.Y. Acad. Sci.* **94**, 668.
Paunier, L. and Rotthauwe, H. W. (1963). *Enzymol. biol. clin.* **3**, 87.
Peacock, A. C., Reed, R. A. and Highsmith, E. M. (1963). *Clinica chim. Acta* **8**, 914.
Pearse, A. G. E. (1957). *J. Histochem. Cytochem.* **5**, 515.
Pearson, C. M., Kar, N. C., Peter, J. P. and Munsat, T. L. (1965). *Am. J. Med.* **39**, 91.
Pelzer, C. F. (1965). *Genetics, Princeton* **52**, 819.
Pesce, A., McKay, R. H., Stolzenbach, F., Cahn, R. D. and Kaplan, N. O. (1964). *J. biol. Chem.* **239**, 1753.
Petras, M. L. (1963). *Proc. natn. Acad. Sci. U.S.A.* **50**, 113.
Pfleiderer, G. (1965). *In* "Enzymes in Clinical Chemistry", p. 105. (R. Ruyssen and L. Vandendriesche, eds.). Elsevier Publishing Company, Amsterdam.
Pfleiderer, G. and Jeckel, D. (1957). *Biochem. Z.* **331**, 103.
Pfleiderer, G. and Wachsmuth, E. D. (1961). *Biochem. Z.* **334**, 185.
Philip, J. and Vesell, E. S. (1962). *Proc. Soc. exp. Biol. Med.* **110**, 582.
Pinter, I. (1957). *Acta physiol. hung.* **11**, 39.

Pinto, P. V. C., Newton, W. A. and Richardson, K. E. (1966). *J. clin. Invest.* **45**, 823.

Plagemann, P. G. W., Gregory, K. F. and Wroblewski, F. (1960a). *J. biol. Chem.* **235**, 2282.

Plagemann, P. G. W., Gregory, K. F. and Wroblewski, F. (1960b). *J. biol. Chem.* **235**, 2288.

Plagemann, P. G. W., Gregory, K. F. and Wroblewski, F. (1961). *Biochem. Z.* **334**, 37.

Plagemann, P. G. W., Gregory, K. F., Swim, H. E. and Chan, K. K. W. (1963). *Can. J. Microbiol.* **9**, 75.

Plummer, D. T. and Wilkinson, J. H. (1963). *Biochem. J.* **87**, 423.

Plummer, D. T., Elliot, B. A., Cooke, K. B. and Wilkinson, J. H. (1963a). *Biochem. J.* **87**, 416.

Plummer, D. T., Wilkinson, J. H. and Withycombe, W. A. (1963b). *Biochem. J.* **89**, 49P.

Polson, A. (1961). *Biochim. biophys. Acta* **50**, 565.

Pope, C. E. and Cooperband, S. R. (1966). *Gastroenterology* **50**, 631.

Popp, R. A. and Popp, D. M. (1962). *J. Hered.* **53**, 111.

Porter, I. H., Boyer, S. H., Watson-Williams, E. J., Adam, A., Szeinberg, A. and Siniscalco, M. (1964). *Lancet* **i**, 895.

Poulik, M. D. (1957). *Nature, Lond.* **180**, 1477.

Poulik, M. D. (1959). *J. Immun.* **82**, 502.

Poulik, M. D. and Smithies, O. (1958). *Biochem. J.* **68**, 636.

Poznanska-Linde, H., Wilkinson, J. H. and Withycombe, W. A. (1966). *Nature, Lond.* **209**, 727.

Press, E. M. and Porter, R. R. (1962). *Biochem. J.* **83**, 172.

Preston, J. A., Batsakis, J. G. and Briere, R. O. (1964). *Am. J. clin. Path.* **41**, 237.

Preston, J. A., Briere, R. O. and Batsakis, J. G. (1965). *Am. J. clin. Path.* **43**, 256.

Prout, G. R., Macalalag, E. V. and Hume, D. M. (1964). *Surgery, St Louis* **56**, 283.

Pyorala, K., Gordin, R., Kontinnen, A. and Telivuo, L. (1963). *Acta Med. Scand.* **174**, 361.

Raacke, I. D. and Li, C. H. (1954). *Biochim. biophys. Acta* **14**, 290.

Racusen, D. and Calvanico, N. (1964). *Analyt. Biochem.* **7**, 62.

Rajewsky, K., Avrameas, S., Grabar, P., Pfleiderer, G. and Wachsmuth, E. D. (1964). *Biochim. biophys. Acta* **92**, 248.

Ramot, B. and Bauminger, S. (1963). *Biochim. biophys. Acta* **73**, 186.

Ramot, B. and Brok, F. (1964). *Ann. hum. Genet.* **28**, 167.

Ramot, B., Bauminger, S., Brok, F., Gafni, D. and Schwartz, J. (1964). *J. Lab. clin. Med.* **64**, 895.

Randerson, S. (1965). *Genetics, Princeton* **52**, 999.

Rasmusen, B. A. (1965). *Genetics, Princeton* **51**, 767.

Raunio, V. and Gabriel, O. (1963). *Nature, Lond.* **197**, 1012.

Raymond, S. (1964). *Ann. N.Y. Acad. Sci.* **121**, 350.

Raymond, S. and Wang, Y-J. (1960). *Analyt. Biochem.* **1**, 391.

Raymond, S. and Weintraub, L. (1959). *Science, N.Y.* **130**, 711.

Reiner, E., Seuforth, W. and Hardegg, W. (1965). *Nature, Lond.* **205**, 1110.

Reisfeld, R. A., Lewis, J. and Williams, D. E. (1962). *Nature, Lond.* **195**, 281.

Reith, A. and Schmidt, F. W. (1964). *Klin. Wschr.* **42**, 915.

Reith, A., Mohr, J., Schmidt, E., Schmidt, F. W. and Wildhirt, E. (1964). *Klin. Wschr.* **42**, 909.

Rendel, J. and Stormont, C. (1964). *Proc. Soc. exp. Biol. Med.* **115**, 853.

Ressler, N., Schulz, J. and Joseph, R. R. (1963a). *J. Lab. clin. Med.* **62**, 571.

Ressler, N., Schulz, J. and Joseph, R. R. (1963b). *Nature, Lond.* **198**, 888.

Ressler, N., Olivero, E., Thompson, G. R. and Joseph, R. R. (1966). *Nature, Lond.* **210**, 695.

Rettenbacher-Daubner, H. and Reider, H. (1963). *Wien. klin. Wschr.* **75**, 833.

Richter, D. and Croft, P. G. (1942). *Biochem. J.* **36**, 746.

Richterich, R. and Burger, A. (1963). *Enzymol. biol. clin.* **3**, 65.

Richterich, R., Schafroth, P. and Aebi, H. (1963). *Clinica chim. Acta* **8**, 178.

Richterich, R., Locher, J., Zuppinger, K. and Rossi, E. (1962). *Schweiz. med. Wschr.* **92**, 919.

Rickli, E., Ghazanfar, A. S., Gibbons, B. H. and Edsall, J. T. (1964). *J. biol. Chem.* **239**, 1065.

Rieder, R. F. and Weatherall, D. J. (1964). *Nature, Lond.* **203**, 1363.

Riggins, R. S. and Kiser, W. S. (1964). *Invest. Urol.* **2**, 30.

Ringoir, S. and Wieme, R. J. (1965). *Lancet Lond.* **ii**, 906.

Roberts, E. (1960). *In* "The Enzymes", Vol. 4, p. 285. (P. D. Boyer, L. Hardy and K. Myrbäck, eds.). Academic Press, New York.

Robb, D. A., Mapson, L. W. and Swain, T. (1965). *Phytochem.* **4**, 731.

Robboy, S. J. and Kahn, R. H. (1964). *Endocrinology* **75**, 97.

Robinson, J. C. and Pierce, J. E. (1964). *Nature, Lond.* **204**, 472.

Robinson, J. C., Pierce, J. E. and Goldstein, D. P. (1965). *Science, N.Y.* **150**, 58.

Robinson, J. C., Pierce, J. E., Goldstein, D. P. and Rosse, W. R. (1966). *Lancet* **ii**, 805.

Robson, E. B. and Harris, H. (1965). *Nature, Lond.* **207**, 1257.

Robson, E. B. and Harris, H. (1966). *Ann. hum. Genet.* **29**, 403.

Roche, J., Thoai, N-v. and Bandoin, J. (1942). *Bull. Soc. Chim. biol. Paris* **24**, 247.

Rosa, J. and Schapira, F. (1965). *Nature, Lond.* **204**, 883.

Rose, R. G. and Wilson, A. C. (1966). *Science, N.Y.* **153**, 1411.

Rosalki, S. B. (1963). *Br. Heart J.* **26**, 795.

Rosalki, S. B. (1965). *Nature, Lond.* **207**, 414.

Rosalki, S. B. and Wilkinson, J. H. (1960). *Nature, Lond.* **188**, 1110.

Rosalki, S. B. and Wilkinson, J. H. (1964). *J. Am. med. Ass.* **189**, 161.

Rosenberg, I. N. (1959). *J. clin. Invest.* **38**, 630.

Rudolph, K. and Stahmann, M. A. (1964). *Nature, Lond.* **204**, 474.

Sacktor, B. (1958). *Proc. IVth Inter. Congress Biochem.* **12**, 138.

Salthe, S. N. (1965). *Comp. Biochem. Physiol.* **16**, 393.

Sandler, M. and Bourne, G. H. (1961). *Exp. Cell Res.* **24**, 174.

Sandler, M. and Bourne, G. H. (1962). *Nature, Lond.* **194**, 389.

Saraswathi, S. and Bachhawat, B. K. (1966). *J. Neurochem.* **13**, 237.

Sawaki, S., Norikawa, N. and Yamada, K. (1965). *Nature, Lond.* **207**, 523.

Sayre, F. W. and Hill, B. R. (1957). *Proc. Soc. exp. Biol. Med.* **96**, 695.

Scandalios, J. G. (1964). *J. Hered.* **55**, 281.

Scandalios, J. G. (1965). *J. Hered.* **66**, 177.

Schapira, F. (1966). *C.r. Acad. Sci. Paris* **262**, 2291.

Schlamowitz, M. (1954a). *J. biol. Chem.* **206**, 361.

Schlamowitz, M. (1954b). *J. biol. Chem.* **206**, 369.

Schlamowitz, M. (1958). *Ann. N.Y. Acad. Sci.* **75**, 373.

Schlamowitz, M. and Bodansky, O. (1959). *J. biol. Chem.* **234**, 1433.

Schleisenger, M. J. and Levinthal, C. (1963). *J. Molec. Biol.* **7**, 1.

Schleuren, P. G. and Goll, R. (1961). *Klin. Wschr.* **39**, 696.

Schneiderman, H., Young, W. J. and Childs, B. (1966). *Science, N.Y.* **151**, 461.

Schobel, B. and Wewalka, F. (1962). *Klin. Wschr.* **40**, 1048.

Schreffler, D. C. (1965). *Am. J. hum. Genet.* **17**, 71.

Schwartz, D. (1960). *Proc. natn. Acad. Sci. U.S.A.* **46**, 1210.

Schwartz, D. (1964a). *Proc. natn. Acad. Sci. U.S.A.* **51**, 602.

Schwartz, D. (1964b). *Proc. natn. Acad. Sci. U.S.A.* **52**, 222.

Schwartz, D. (1965). *Genetics, St. Louis* **52**, 1295.

Schwartz, D. and Endo, T. (1966). *Genetics, St. Louis* **53**, 709.

Schwartz, D., Fuchsman, L. and McGrath, K. H. (1965). *Genetics, Lond.* **52**, 1265.

Schwartz, H. M., Biedron, S. I., Van Holdt, M. M. and Rehm, S. (1964). *Phytochem.* **3**, 189.

Schwartz, M. K., Nisselbaum, J. S. and Bodansky, O. (1963). *Am. J. clin. Path.* **40**, 103.

Scott, E. M. (1966). *J. biol. Chem.* **241**, 3049.

Scott, E. M., Duncan, I. W., Ekstrand, V. and Wright, R. C. (1966). *Am. J. hum. Genet.* **18**, 408.

Searcy, R. L., Hayashi, S., Hardy, E. M. and Berk, J. E. (1965). *Clinica chim. Acta* **12**, 631.

Searcy, R. L., Ujihira, I., Hayashi, S. and Berk, J. E. (1964). *Clinica chim. Acta* **9**, 505.

Secchi, G. C. and Dioguardi, N. (1965). *Enzymol. biol. clin.* **5**, 29.

Secchi, G. C., Mossa, R. and Gallitelli, L. (1964). *Enzymol. biol. clin.* **4**, 58.

Secchi, G. C., Rezzonico, A. and Gervasini, N. (1965). *Atti. Accad. med. lomb.* **XX**, 64.

Sen, M., Drance, S. M. and Woodford, V. R. (1963). *Can. J. Biochem. Physiol.* **41**, 1235.

Shaw, C. R. (1965). *Science, N.Y.* **149**, 936.

Shaw, C. R. (1966). *Science, N.Y.* **153**, 1013.

Shaw, C. R. and Barto, E. (1963). *Proc. natn. Acad. Sci. U.S.A.* **50**, 211.

Shaw, C. R. and Barto, E. (1965). *Science, N.Y.* **148**, 1099.

Shaw, C. R. and Koen, A. L. (1963). *Science, N.Y.* **140**, 70.

Shaw, C. R. and Koen, A. L. (1965). *J. Histochem. Cytochem.* **13**, 431.

Shaw, C. R., Syner, F. N. and Tashian, R. E. (1962). *Science, N.Y.* **138**, 31.

Sheid, B. and Roth, J. S. (1965). *In* "Advances in Enzyme Regulation" Vol. III, p. 335 (G. Weber, ed.), Pergamon Press, Oxford.

Sheid, B., Morris, H. P. and Roth, J. S. (1965). *J. biol. Chem.* **240**, 3016.

Shulman, S. and Ferber, J. M. (1966). *J. Reprod. Fert.* **11**, 295.

Shulman, S., Mamrod, L., Gonder, M. J. and Soanes, W. H. (1964). *J. Immun.* **93**, 474.

Sick, K. and Nielson, J. T. (1964). *Hereditas* **51**, 291.

Smith, C. H. and Kissane, J. M. (1965). *Dev. Biol.* **8**, 151.

Smith, E. E. and Rutenberg, A. M. (1963). *Nature, Lond.* **197**, 800.

Smith, E. E. and Rutenberg, A. M. (1966). *Science, Lond.* **152**, 1256.

Smith, E. E., Pineda, E. P. and Rutenberg, A. M. (1962). *Proc. Soc. exp. Biol. Med.* **110**, 683.

Smith, J. L. and Krueger, R. C. (1962). *J. biol. chem.* **237**, 1121.

Smith, K. D., Ursprung, H. and Wright, T. R. F. (1963). *Science, N.Y.* **142**, 226.

Smith, L. C., Ravel, J. M., Lax, S. R. and Shive, W. (1962). *J. biol. Chem.* **237**, 2566.

Smithies, O. (1955). *Biochem. J.* **61**, 629.

Smithies, O. (1959a). *Biochem. J.* **71**, 585.

Smithies, O. (1959b). *In* "Advances in Protein Chemistry", Vol. 14, p. 65. (M. L. Anson and T. E. Edsall, eds.). Academic Press New York.

Smithies, O. (1963). *Archs Biochem. Biophys.* Supp. I, 125.

Smithies, O. and Dixon, G. H. (1957). *Biochim. biophys. Acta* **23**, 198.

Sober, H. A. and Peterson, E. A. (1954). *J. Am. chem. Soc.* **76**, 1711.

Sober, H. A., Gutter, F. J., Syckoff, M. M. and Peterson, E. A. (1956). *J. Am. chem. Soc.* **78**, 756.

Soetens, A., Karcher, D., Van Sande, M. and Lowenthal, A. (1965). *In* "Enzymes in Clinical Chemistry", p. 130. (R. Ruyssen and L. Vandendriessche, eds.). Elsevier Publishing Co., Amsterdam.

Spencer, N., Hopkinson, D. A. and Harris, H. (1964). *Nature, Lond.* **204**, 742.

Stadtman, E. R. (1963). *Bact. Rev.* **27**, 170.

Stadtman, E. R., Cohen, G. N. and LeBras, G. (1961). *Ann. N.Y. Acad. Sci.* **94**, 952.

Stambaugh, R. and Post, D. (1966a). *J. biol. Chem.* **241**, 1462.

Stambaugh, R. and Post, D. (1966b). *Biochim. biophys. Acta* **122**, 541.

Stambaugh, R. and Post, D. (1966c). *Ann. Biochem.* **15**, 470.

Stanbury, J. B. (1957). *J. biol. Chem.* **228**, 801.

Staples, R. C. and Stahmann, M. A. (1964). *Phytopathology* **54**, 760.

Staples, R. C., McCarthy, W. J. and Stahmann, M. A. (1965). *Science, N.Y.* **149**, 1248.

Starkweather, W. H. and Schoch, H. K. (1962). *Biochim. biophys. Acta* **62**, 440.

Starkweather, W. H., Cousineau, L., Schoch, H. K. and Zaraforetis, C. J. (1965). *Blood* **26**, 63.

Starkweather, W. H., Green, R. A., Schwartz, E. L. and Schoch, H. K. (1966a). *J. Lab. clin. Med.* **67**, 329.

Starkweather, W. H., Green, R. A., Spencer, H. H. and Schoch, H. K. (1966b). *J. Lab. clin. Med.* **68**, 314.

Stern, J. and Lewis, W. H. P. (1962). *J. ment. Defic. Res.* **6**, 13.

Stevenson, D. E. (1961). *Clinica chim. Acta* **6**, 142.

Strandjord, P. E. and Clayson, K. J. (1961). *J. Lab. clin. Med.* **58**, 962.

Strandjord, P. E., Clayson, K. J. and Freier, E. F. (1962). *J. Am. med. Ass.* **182**, 1099.

Suld, H. M. and Herbut, P. A. (1965). *J. biol. Chem.* **240**, 2234.

Sundaram, T. K. and Fincham, J. R. S. (1964). *J. molec. Biol.* **10**, 423.

Sur, B. K., Moss, D. W. and King, E. J. (1962). *Proc. Ass. clin. Biochem.* **2**, 11.

Svensmark, O. (1961a). *Dan. med. Bull.* **8**, 28.

Svensmark, O. (1961b). *Acta physiol. scand.* **52**, 372.

Swick, R. W., Barnstein, P. L. and Stange, J. L. (1965). *J. biol. Chem.* **240**, 3334.

Syner, F. N. and Goodman, M. (1966). *Science, N.Y.* **151**, 206.

Takahashi, H., Wazima, T. and Mizushima, A. (1963). Abstracted in *Excerpta med.* **17**, Section II, No. 1636, (1964).

Takasu, T. and Hughes, B. P. (1966). *Nature, Lond.* **212**, 609.

Talal, N., Hermann, G. and De Vaux St. Cyr, C. (1963). Protides of the Biological Fluids, p. 183. 10th Colloq. Bruges, Elsevier Publishing Co., Amsterdam.

Tappan, D. V., Jacey, M. J. and Boyden, M. (1964). *Ann. N.Y. Acad. Sci.* **121**, 589.

Tashian, R. E. (1961). *Proc. Soc. exp. Biol. Med.* **108**, 364.

Tashian, R. E. (1965). Personal communication to E. S. Vessel (1965c).

Tashian, R. E. and Shaw, M. W. (1962). *Am. J. hum. Genet.* **14**, 295.

Tashian, R. E., Plato, C. C. and Shows, T. B. (1963). *Science, N.Y.* **140**, 53.

Taswell, H. F. and Jeffers, D. M. (1963). *Am. J. clin. Path.* **40**, 349.

Thiele, K. G. and Mattenheimer, H. (1966). *Z. klin. Chem.* **4**, 232.

Thompson, R. R. and Cook, J. W. (1961). *J. Ass. off. agric. Chem.* **44**, 199.

Thorne, C. J. R., Grossman, L. I. and Kaplan, N. O. (1963). *Biochim. biophys. Acta* **73**, 193.

Thorup, O. A., Carpenter, J. T. and Howard, P. (1964). *Br. J. Haemat.* **10**, 542.

Thorup, O. A., Stole, W. B. and Leavell, B. S. (1961). *J. Lab. clin. Med.* **58**, 122.

Thurman, D. A., Palin, C. and Laycock, M. V. (1965). *Nature, Lond.* **207**, 193.

Trubowitz, S. and Miller, W. L. (1966). *Proc. Soc. exp. Biol. Med.* **123**, 187.

Trujillo, J. M., Walden, B., O'Neill, P. and Anstall, H. B. (1965). *Science, N.Y.* **148**, 1603.

Tsao, M. U. (1960). *Archs Biochem. Biophys.* **90**, 234.

Turner, D. M. (1964). *Proc. Ass. clin. Biochem.* **3**, 14.

Ujihira, I., Searcy, R. L., Berk, J. E. and Hayashi, S. (1965). *Clin. Chem.* **11**, 97.
Ulrich, F. (1964). *Int. Cong. Biochem.* 6(4), 340.
Umbarger, H. E. and Brown, B. (1957). *J. Bact.* **73**, 105.
Umbarger, H. E. and Brown, B. (1958). *J. biol. Chem.* **233**, 1156.
Unjehm, O., From, S. H. J. and Schulz-Haudt, S. D. (1966). *Acta chem. scand.* **20**, 799.
Uriel, J. (1957). *Bull. Soc. Chim. biol.* Supp. 1, 105.
Uriel, J. (1960). *Nature, Lond.* **188**, 853.
Uriel, J. (1961). *Annls Inst. Pasteur, Paris* **101**, 104.
Uriel, J. (1963). *Ann. N.Y. Acad. Sci.* **103**, 956.
Ursprung, H. and Leone, J. (1965). *J. exp. Zool.* **160**, 147.
Van Asperen, K. (1962). *J. Insect Physiol.* **8**, 401.
Van Asperen, K. (1964). *Entomologia exp. appl.* **1**, 205.
Van Asperen, K. and Van Mazijk, M. (1965). *Nature, Lond.* **205**, 1291.
Van der Helm, H. J. (1961). *Lancet* **ii**, 108.
Van der Helm, H. J. (1962b). *Nature, Lond.* **194**, 773.
Van der Helm, H. J. (1962a). *J. Neurochem.* **9**, 325.
Van der Helm, H. J., Zondag, H. A. and Klein, F. (1963). *Clinica chim. Acta* **8**, 193.
Van der Helm, H. J., Zondag, H. A., Hartog, A. Ph. and Van der Kooi, M. W. (1962). *Clinica chim. Acta* **7**, 540.
Van der Jooste, J. and Morland, D. E. (1963). *Phytochem.* **2**, 263.
Van der Veen, K. J. and Willebrands, A. F. (1966). *Clinica chim. Acta* **13**, 312.
Van Ros, G. and Druet, R. (1966). *Nature, Lond.* **212**, 543.
Van Wijhe, M., Blanchaer, M. C. and St. George Stubbs, S. (1964). *J. Histochem. Cytochem.* **12**, 608.
Vecchio, F., Schettini, F., Di Francesco, L., Meloni, T. and Russino, G. (1966). *Acta Haemat.* **35**, 46.
Velthius, H. H. W. and Van Asperen, K. (1963). *J. Histochem. Cytochem.* **6**, 79.
Vesell, E. S. (1961). *Ann. N.Y. Acad. Sci.* **94**, 877.
Vesell, E. S. (1962). *Nature, Lond.* **195**, 497.
Vesell, E. S. (1964). *In* "Protides of the Biological Fluids", p. 510. (H. Peeters, ed.). Elsevier Publishing Company, Amsterdam.
Vesell, E. S. (1965d). *Science, N.Y.* **148**, 1103.
Vesell, E. S. (1965b). "Genetic control of isoenzyme patterns in human tissues", *in* Progress in Medical Genetics. Vol. IV, p. 128. (A. G. Steinberg and A. G. Bearn, eds.). Wm. Heinemann, London.
Vesell, E. S. (1965c). *Science, N.Y.* **150**, 1590.
Vesell, E. S. (1965a). *Science, N.Y.* **150**, 1735.
Vesell, E. S. (1966). *Nature, Lond.* **210**, 421.
Vesell, E. S. and Bearn, A. G. (1957). *Proc. Soc. exp. Biol. Med.* **94**, 96.
Vesell, E. S. and Bearn, A. G. (1958). *J. clin. Invest.* **37**, 672.
Vesell, E. S. and Bearn, A. G. (1961). *J. clin. Invest.* **40**, 586.
Vesell, E. S. and Bearn, A. G. (1962). *J. gen. Physiol.* **45**, 553.
Vesell, E. S. and Brody, I. A. (1964). *Ann. N.Y. Acad. Sci.* **121**, 544.
Vesell, E. S. and Pool, P. E. (1966). *Proc. natn. Acad. Sci. U.S.A.* **55**, 756.
Vesell, E. S., Feldman, M. P. and Frank, E. D. (1959). *Proc. exp. Biol. Med.* **101**, 644.
Vesell, E. S., Philip, J. and Bearn, A. G. (1962a). *J. exp. Med.* **116**, 797.
Vesell, E. S., Osterland, K. C., Bearn, A. G. and Kunkel, H. (1962b). *J. clin. Invest.* **41**, 2012.
Von Wartburg, J. P., Papenburg, J. and Aebi, H. (1965). *Can. J. Biochem. Physiol.* **43**, 889.

Wachsmuth, E. D. (1964). *Nature, Lond.* **204**, 681.
Wachsmuth, E. D. and Pfleiderer, G. (1963). *Biochem. Z.* **336**, 545.
Wachsmuth, E. D., Pfleiderer, G. and Wieland, Th. (1964). *Biochem. Z.* **340**, 80.
Wacker, W. E. C. and Schoenenberger, G. A. (1966). *Biochem. biophys. Res. Commun.* **22**, 291.
Wada, H. and Morino, Y. (1964). *Vitamns Horm.* **22**, 411.
Walker, J. R. L. and Hulme, A. C. (1965). *Phytochem.* **5**, 259.
Walsh, K. A., Ericcson, L. H. and Neurath, H. (1966). *Proc. natn. Acad. Sci. U.S.A.* **56**, 1339.
Walter, H. and Selby, F. W. (1966). *Nature, Lond.* **212**, 613.
Walter, H., Selby, F. W. and Francisco, J. R. (1965). *Nature, Lond.* **208**, 76.
Warburton, F. G. (1965) in Wilkinson (1965a). 62.
Warburton, F. G., Smith, D. and Laing, G. S. (1963). *Nature, Lond.* **198**, 386.
Warnock, M. L. (1964). *Proc. Soc. exp. Biol. Med.* **115**, 448.
Warnock, M. L. (1966). *Clinica chim. Acta* **14**, 156.
Watts, D. C. and Donniger, C. (1962). *Analyt. Biochem.* **3**, 489.
Weber, D. J. and Stahmann, M. A. (1964). *Science, N.Y.* **146**, 929.
Weber, G. and Pfleiderer, G. (1961). *Ann. N.Y. Acad. Sci.* **94**, 933.
Weiser, M. M., Bolt, R. J. and Pollard, H. N. M. (1964). *J. Lab. clin. Med.* **63**, 656.
White, J. W. and Kushnir, I. (1966). *Analyt. Biochem.* **16**, 302.
Wieland, Th. and Pfleiderer, G. (1957). *Biochem. Z.* **329**, 112.
Wieland, Th. and Pfleiderer, G. (1961). *Ann. N.Y. Acad. Sci.* **94**, 691.
Wieland, I. and Pfleiderer, G. (1962). *Angew. Chem.* (International Edition) **1**, 169.
Wieland, I., Duesberg, P. and Determann, H. (1963). *Biochem. Z.* **337**, 303.
Wieland, I., Pfleiderer, G. and Ortanderl, F. (1959b). *Biochem. Z.* **331**, 103.
Wieland, I., Pfleiderer, G. and Rajewsky, K. (1960). *Naturforsch.* **15b**, 434.
Wieland, I., Pfleiderer, G., Haupt, I. and Worner, W. (1959a). *Biochem. Z.* **332**, 1.
Wieme, R. J. (1958). Protides of the Biological Fluids. 6th Colloquim. Bruges. Elsevier Publishing Co., Amsterdam.
Wieme, R. J. (1959a). *Clinica chim. Acta* **4**, 46.
Wieme, R. J. (1959b). *Clinica chim. Acta* **4**, 317.
Wieme, R. J. (1963). *Nature, Lond.* **199**, 437.
Wieme, R. J. (1966). *Clinica chim. Acta* **13**, 138.
Wieme, R. J. (1962). *Nature, Lond.* **194**, 287.
Wieme, R. J. and Herpol, J. E. (1962). *Nature, Lond.* **194**, 284.
Wieme, R. J. and Lauryssens, M. J. (1962). *Lancet* **i**, 433.
Wieme, R. J. and Van Maercke, Y. (1961). *Ann. N.Y. Acad. Sci.* **94**, 898.
Wiggert, B. and Villee, C. A. (1964). *J. biol. Chem.* **239**, 444.
Wilde, C. E. and Kekwick, R. G. O. (1964). *Biochem. J.* **91**, 297.
Wilding, P. (1963). *Clinica chim. Acta* **8**, 918.
Wilding, P. (1965). *Clinica chim. Acta* **12**, 97.
Wilding, P., Cooke, W. T. and Nicholson, G. I. (1964). *Ann. intern. Med.* **60**, 1053.
Wilkinson, J. H. (1965a). "Isoenzymes". Spon. London.
Wilkinson, J. H. (1965b). *Geriatrics* **20**, 637.
Wilkinson, J. H. and Rosalki, S. B. (1963). *Diag Terap.* **1**, 309.
Wilkinson, J. H. and Withycombe, W. A. (1965). *Biochem. J.* **97**, 663.
Wilkinson, J. H., Cooke, K. B., Elliot, B. A. and Plummer, D. T. (1961). *Biochem. J.* **80**, 29P.
Wilkinson, J. H., Elliot, B. A., Cooke, K. B. and Plummer, D. T. (1962). *Proc. Ass. clin. biochem.* **2**, 1.
Wilson, A. C., Cahn, R. D. and Kaplan, N. O. (1963). *Nature, Lond.* **197**, 331.
Wilson, A. C., Kaplan, N. O., Levine, L., Pesce, A., Reichlin, M. and Allison, W. S. (1964). *Fedn Proc. Fedn Am. Socs. exp. Biol.* **23**, 1258.
Withycombe, W. A., Plummer, D. T. and Wilkinson, J. H. (1965). *Biochem. J.* **94**, 384.

Witt, I., Kronau, R. and Holzer, H. (1966). *Biochim. Biophys. Acta* **128**, 63.

Woerner, W. and Martin, H. (1961). *Klin. Wschr.* **39**, 368.

Woerner, W. and Martin, H. (1962). Proc. 8th Congress of European Soc. Haemat. Art. 26, Vienna, 1961.

Wong, P. W. K., Shih, L. Y., Hsia, D. Y-Y, and Tsao, Y. C. (1965). *Nature, Lond.* **208**, 1323.

Wood, T. (1963). *Biochem. J.* **87**, 453.

Wright, T. R. F. (1963). *Genetics, Princeton* **48**, 787.

Wright, T. R. F. and McIntyre, R. J. (1963). *Genetics, Princeton* **48**, 1717.

Wroblewski, F. (1963). *Prog. Cardiovasc. Dis.* **6**, 63.

Wroblewski, F. and Gregory, K. (1961). *Ann. N.Y. Acad. Sci.* **94**, 912.

Wroblewski, F., Ross, C. and Gregory, K. (1960). *New Engl. J. med.* **263**, 531.

Wüst, H., Schön, H. and Berg, G. (1962). *Klin. Wschr.* **40**, 1169.

Yakulis, V. J., Gibson, C. W. and Heller, P. (1962). *Am. J. clin. Path.* **38**, 378.

Yasin, R. and Goldenberg, G. J. (1966). *Nature, Lond.* **211**, 1296.

Yen, T. T. T. and Glassman, E. (1965). *Genetics, Princeton* **52**, 977.

Yong, J. M. (1966). *Lancet* **i**, 1132.

Young, W. J., Porter, J. E. and Childs, B. (1964). *Science, N.Y.* **143**, 140.

Zinkham, W. H., Blanco, A. and Kupchyk, L. (1964a). *Science, N.Y.* **144**, 1353.

Zinkham, W. H., Blanco, A. and Clowry, L. J. (1964b). *Ann. N.Y. Acad. Sci.* **121**, 571.

Zinkham, W. H., Blanco, A. and Kupchyk, L. (1963). *Science, N.Y.* **142**, 1303.

Zinkham, W. H., Blanco, A. and Kupchyk, L. (1966b). *Pediatrics, N.Y.* **37**, 120.

Zinkham, W. H., Kupchyk, L., Blanco, A. and Isensee, H. (1966a). *J. exp. Zool.* **162**, 45.

Zondag, H. A. (1963). *Science, N.Y.* **142**, 965.

Zondag, H. A. (1965). "Enzymes in Clinical Chemistry", pp. 120–129. (R. Ruyssen and L. Vandendriessche, eds.). Elsevier Publishing Co., Amsterdam.

Appendix

The following list of literature references with titles has been collected since the literature survey for this monograph was completed, and is arranged in order of the appropriate chapters.

CHAPTER II

Anderson, S. and Weber, G. (1966). The reversible acid dissociation and hybridization of lactic dehydrogenase. *Archs Biochem. Biophys.* **116**, 207.

Arnheim, N. Jr., Cocks, G. T. and Wilson, A. C. (1967). Molecular size of hagfish muscle lactate dehydrogenase. *Science, N.Y.* **157**, 568.

Atanasov, N. (1966). Apropos of the determination of the total lactate dehydrogenase and its isoenzymes. *Folia Med. (Plovdiv)* **8**, 126.

Balek, R. W., Snow, J. and Haduck, L. (1967). Electrophoresis and heat inactivation of lactate dehydrogenase isozymes in amphibia. *Life Sciences* **6**, 1035.

Bernstein, L., Kerrigan, M. and Maisel, H. (1966). Lactic dehydrogenase isozymes in lens and cornea. *Expl. Eye Res.* **5**, 309.

Blackshaw, A. W. and Samisoni, J. I. (1966). The effects of cryptorchism in the guinea pig on the isoenzymes of testicular lactate dehydrogenase. *Aust. J. Biol. Sci.* **19**, 841.

Blanco, A., Rife, U. and Larson, B. L. (1967). Lactate dehydrogenase isozymes during dedifferentiation in cultures of mammary secretory cells. *Nature, Lond.* **214**, 1331.

Blatt, W. F., Walker, J. and Mager, M. (1965). Tissue lactic dehydrogenase isozymes: variation in rat during prolonged cold exposure. *Am. J. Physiol.* **209**, 785.

Blonde, D. J., Kresack, E. J. and Kosicki, G. W. (1967). Effects of ions and freeze-thawing in supernatant and mitochondrial malate dehydrogenase. *Can. J. Biochem. Physiol.* **45**, 641.

Bloom, A. D., Tsuchioka, M. and Wajima, T. (1967). Lactic dehydrogenase and metabolism of human leukocytes in vitro. *Science, N.Y.* **156**, 979.

Bohn, L. (1966). Laktatdehydrogenasens isoenzymer. *Nord Méd.* **76**, 1248.

Bonavita, V., Amore, G. and Avellone, S. (1966). Molecular and kinetic properties of lactate dehydrogenase in the degenerating peripheral nerve. *J. Neurol. Sci.* **3**, 340.

Boyd, J. W. (1967). The rates of disappearance of L-Lactate dehydrogenase isoenzymes from plasma. *Biochim. biophys. Acta* **132**, 221.

Buckley, R. D. and Balchum, D. J. (1967). Effects of nitrogen dioxide on lactic dehydrogenase isozymes. *Archs. Envir. Hlth* **14**, 424.

Dioguardi, N., Ideo, G., Mannucci, P. M., Fiorelli, G. and Agostoni, A. (1966). Multiple molecular forms of lactate dehydrogenase (LDH) of normal and leukemic cells of the myeloid line. *Enzymol. biol. clin.* **6**, 1.

Dioguardi, N., Ideo, G., Mannucci, P. M. and Fiorelli, G. (1966). Lactate dehydrogenase (LDH) glutamic oxalacetic transaminase (GOT) and malate dehydrogenase (MDH) isoenzymes in lymphocytes from foetal and adult thymus, spleen and from peripheral blood. *Enzymol. biol. clin.* **6**, 324.

Di Sabato, G. (1966). The hydrogen ion equilibria of chicken heart lactic dehydrogenase. *Biochemistry, N.Y.* **5**, 3980.

Doman, E. and Koide, S. S. (1966). Analysis of 3-alpha and 3-beta-hydroxysteroid oxidoreductases of rat liver by disc electrophoresis. *Biochim. biophys. Acta* **128**, 209.

Fritz, P. J. (1967). Rabbit lactate dehydrogenase isozymes: effect of pH on activity. *Science, N.Y.* **156**, 82.

Gordon, H., Keraan, M. M. and Vooijs, M. (1967). Variants of 6-phosphogluconate dehydrogenase within a community. *Nature, Lond.* **214**, 466.

Gregory, K. F., Ng, C. W. and Pantekoek, J. F. (1966). Antibody to lactate dehydrogenase. 1. Inhibition of glycolysis in tumour and liver homogenates. *Biochim. biophys. Acta* **130**, 469.

Hawrylewicz, E. J. and Blair, W. H. (1966). Effect of gamma and proton irradiation on lactic dehydrogenase isoenzymes. *Radiat. Res.* **28**, 538.

Hawrylewicz, E. J. and Blair, W. H. (1967). Enzyme-isoenzyme measure of radiation exposure. *Aerospace Med.* **38**, 30.

Hornung, G., Lehmann, F. G. and Braun, H. (1967). Der Einfluss der Rontgen-Ganzkorperbestrahlung auf die Enzymaktiven der Laktat-Dehydrogenase-Isoenzyme in verschiedenen Organen und im Blutserum bei Ratten. *Strahlentherapie* **128**, 595.

Houssais, J. F. (1966). Molecular transformations of mouse lactate dehydrogenase isozymes indicated by starch-gel electrophoresis. *Biochim. biophys. Acta* **128**, 239.

Kaufman, L. V. and Jacobson, K. B. (1966). Use of lactate dehydrogenase as a marker in rat-mouse chimera studies. *Cancer Res.* **26**, 1778.

Kitto, G. B. and Kaplan, N. O. (1966). Purification and properties of chicken heart mitochondrial and supernatant malic dehydrogenases. *Biochemistry, N.Y.* **5**, 3966.

Kitto, G. B. and Wilson, A. C. (1966). Evolution of malate dehydrogenase in birds. *Science, N.Y.* **153**, 1408.

Kitto, G. B., Wassarman, P. M. and Kaplan, N. O. (1966). Enzymatically active conformers of mitochondrial malate dehydrogenase. *Proc. natn. Acad. Sci. U.S.A.* **56**, 578.

Kitto, G. B., Wassarman, P. M., Michjeda, J. and Kaplan, N. O. (1966). Multiple forms of mitochondrial malate dehydrogenase. *Biochem. biophys. Res. Commun.* **22**, 75.

Kjellberg, B. and Karlsson, B. (1966). A comparison of changes in lactate dehydrogenase isoenzyme patterns of porcine, human and bovine milk during lactation periods. *Biochim. biophys. Acta* **128**, 589.

Kjellberg, B. and Karlsson, B. W. (1967). Comparative analyses of lactic and malic dehydrogenases and their multiple molecular forms in milk from various animal species and man. *Comp. Biochem. Physiol.* **22**, 397.

Koen, A. L. (1967). Lactate dehydrogenase isozyme sub-bands: new bands derived from isolated sub-bands. *Biochim. biophys. Acta* **140**, 496.

Koen, A. L. and Shaw, C. R. (1965). Studies on lactate dehydrogenase tissue-specific isozymes: electrophoretic migration of isolated sub-bands. *Biochim. biophys. Acta* **96**, 231.

Koen, A. L. and Shaw, C. R. (1966). Retinol and alcohol dehydrogenases in retina and liver. *Biochim. biophys. Acta* **128**, 48.

Lewis, C., Schmitt, M. and Hershey, F. B. (1967). Heterogeneity of lactic dehydrogenase of human skin. *J. invest. Derm.* **48**, 221.

Nutter, D. O., Trujillo, N. P. and Evans, J. M. (1966). The isoenzymes of lactic dehydrogenase. *Am. Heart J.* **72**, 315.

Lojda, Z. and Fric, P. (1966). Lactic dehydrogenase isoenzymes in the aortic wall. *J. Atheroscler. Res.* **6**, 264.

Macalalag, E. N. and Prout, G. R. (1965). Lactate dehydrogenase, its isozymes and the canine prostate. *Invest. Urol.* **3**, 268.

Maisel, H. and Kerrigan, K. (1966). Effect of temperature (and substrate concentration) on chick lactate dehydrogenase activity. *Proc. Soc. exp. Biol. Med.* **123**, 847.

Markert, C. L. and Sladen, W. J. (1966). Stability of lactate dehydrogenase isozyme patterns in penguins. *Nature, Lond.* **210**, 948.

Ng, C. W. and Gregory, K. F. (1966). Antibody to lactate dehydrogenase. 11. Inhibition of glycolysis and growth of tumour cells. *Biochim. biophys. Acta* **130**, 477.

Odense, O., Allen, T. M. and Leung, T. C. (1966). Multiple forms of lactate dehydrogenase and aspartate aminotransferase in herring. *Canad. J. Biochem. Physiol.* **44**, 1319.

Ohkawara, A., Halprin, J. and Halprin, K. M. (1967). Human epidermal isoenzymes. *Arch. Derm.* **95**, 412.

Paulsen, G. D., Pope, A. L. and Baumann, C. A. (1966). Lactic dehydrogenase isoenzymes in tissues and serum of normal and dystrophic lambs. *Proc. Soc. exp. Biol. Med.* **122**, 321.

Pesce, A., Fondy, T. P., Stolzenbach, F., Castillo, F. and Kaplan, N. O. (1967). The comparative enzymology of lactic dehydrogenases. 3. Properties of the H4 and M4 enzymes from a number of vertebrates. *J. biol. Chem.* **242**, 2151.

Quatrini, U. and Cajola, G. (1965). Lactic dehydrogenase isoenzymes in the blood serum of pigeons with experimental beri-beri. *Boll. Soc. ital. Biol. sper.* **41**, 292.

Rabinowitz, Y. and Dietz, A. A. (1967). Malic and lactic dehydrogenase isozymes of normal and leukemic leukocytes separated on glass bead columns. *Blood* **29**, 182.

Rajewsky, K. and Müller, B. (1967). Similar surface areas on acetylated lactic dehydrogenases. *Immunochemistry* **4**, 151.

Ressler, N. and Tuttle, C. (1966). Significance of sub-bands of lactic dehydrogenase isozymes. *Nature, Lond.* **210**, 1268.

Ressler, N., Olivero, E. and Joseph, R. R. (1965). Lactic dehydrogenase isozymes in human testis. *Nature, Lond.* **206**, 829.

Ressler, N., Stitzer, K. L. and Ackerman, D. R. (1967). Reactions of the lactate dehydrogenase X-band in human sperm with homologous and heterologous antisera. *Biochim. biophys. Acta* **139**, 507.

Roddick, J. W. Jr., Ing, G. K. and Midboe, D. (1966). Isozymes of lactic dehydrogenase in normal endometrium. *Am. J. Obstet. Gynec.* **95**, 459.

Schmidt, E., Schmidt, F. W., Herfarth, C., Opitz, K. and Vogell, W. (1966). Studies of the efflux of enzymes from the model of the isolated perfused rat liver. 3. Analysis of the extracellular enzyme patterns after perfusion in hypoxia. *Enzym. biol. clin.* **7**, 185.

Stein, A. M. and Stein, J. H. (1965). Studies on the Straub diaphorase: isolation of multiple forms. *Biochemistry, N.Y.* **4**, 1491.

Stewart, J. A. and Papaconstantinou, J. (1966). Lactate dehydrogenase isozymes and their relationship to lens cell differentiation. *Biochem. biophys. Acta* **121**, 69.

Szeinberg, A., Mor, A., Vernia, H. and Reischer, S. (1966). 'Band-X' isozyme of lactic dehydrogenase in pathological spermatogenesis. *Life Sciences* **5**, 1233.

Thorup, O. A. Jr. (1967). The isozymes of human erythrocyte catalase. *Trans Am. clin. clim. Ass.* **78**, 129.

Trujillo, N. P. and Evans, J. M. (1966). Lactic dehydrogenase isoenzyme. V. A comparison of its determination by heat fractionation and by alpha-hydroxy butyrate dehydrogenase. *Am. J. med. Sci.* **252**, 159.

Trujillo, N. P., Nutter, D. and Evans, J. M. (1967). The isoenzymes of lactic dehydrogenase. *Archs. intern. Med.* **119**, 333.

Van Bogaert, E. C., De Peretti, E. and Villee, C. A. (1967). Electrophoretic studies of human placental dehydrogenases. *Am. J. Obstet. Gynec.* **98**, 919.

Vesell, E. S. (1965). Formation of human lactate dehydrogenase isoenzymes in vitro. *Proc. natn. Acad. Sci. U.S.A.* **54**, 111.

Vesell, E. S. and Yieding, K. L. (1966). Effects of pH, ionic strength, and metabolic intermediates on the rates of heat inactivation of lactate dehydrogenase isozymes. *Proc. natn. Acad. Sci. U.S.A.* **56**, 1317.

Wacker, W. E. and Schoenenberger, G. A. (1966). Peptide inhibitors of lactic dehydrogenase (LDH) I: specific inhibition of LDH-M-4 and LDH-H-4 by inhibitor peptides I and II. *Biochem. biophys. Res. Commun.* **22**, 291.

Walter, H. and Caccan, J. F. (1966). Effect of oxidized glutathione on some enzymes of erythrocytes and its relation to erythrocytic enzyme activity and electrophoretic mobility. *Biochem. J.* **100**, 274.

Wilson, A. C., Kitto, G. B. and Kaplan, N. O. (1967). Enzymatic identification of fish products. *Science, N.Y.* **157**, 82.

Withycombe, W. (1967). Lactate dehydrogenase isoenzymes in chick tissues. *Nature, Lond.* **213**, 513.

Wust, H. (1965). Serum isoenzymes of lactate dehydrogenase in patients with malignant tumors. *Verh. Deitscj ges. Inn. Med.* **71**, 949.

CHAPTER III

Brody, I. A. and Hatcher, M. A. (1967). Origin of increased serum creatine phosphokinase in tetanus. An isoenzyme analysis. *Archs Neurol.* **16**, 89.

Brown, J., Miller, D. M., Holloway, M. T. and Leve, G. D. (1967). Hexokinase isoenzymes in liver and adipose tissue of man and dog. *Science, N.Y.* **155**, 205.

Craig, F. A. and Smith, J. C. (1967). Creatine phosphokinase in thyroid: isoenzyme composition compared with other tissues. *Science, N.Y.* **156**, 254.

Dawson, D. M. and Fine, I. H. (1967). Creatine kinase in human tissues. *Archs Neurol.* **16**, 175.

Dawson, D. M., Eppenberger, H. M. and Kaplan, N. O. (1967). The comparative enzymology of creatine kinases II. *J. biol. Chem.* **242**, 210.

Eppenberger, H. M., Dawson, D. M. and Kaplan, N. O. (1967). The comparative enzymology of creatine kinases I. *J. biol. Chem.* **242**, 204.

Grossbard, L., Weksler, M. and Schimke, R. T. (1966). Electrophoretic properties and tissue distribution of multiple forms of hexokinase in various mammalian species. *Biochem. Biophys. Res. Commun.* **24**, 32.

Hanson, T. L. and Fromm, H. J. (1967). Rat skeletal muscle hexokinase. 11. Kinetic evidence for a second hexokinase in muscle tissue. *J. biol. Chem.* **242**, 501.

Holmes, E. W. Jr., Malone, J. I., Winegrad, A. I. and Oski, F. A. (1967). Hexokinase isoenzymes in human erythrocytes: association of type II with fetal hemoglobin. *Science, N.Y.* **156**, 646.

Hooton, B. T. and Watts, D. C. (1966). Adenosine 5-triphosphate-creatine phosphotransferase from dystrophic mouse skeletal muscle. A genetic lesion associated with the catalytic-site thiol group. *Biochem. J.* **100**, 637.

Kormendy, L. and Gantner, G. (1965). Transaminase-isozyme des Skeletmuskel. *Naturwissenschaften* **52**, 209.

Kormendy, L., Gantner, G. and Hamm, R. (1965). Isozyme der Glutamat-Oxalacetat-Transaminase in Skeletmuskel von Schwein und Rind. *Biochem. J.* **342**, 31.

Mannucci, P. M. and Dioguardi, N. (1966). Electrophoretic characterization of glutamic oxalacetic transaminase in human red cells of different ages. *Clinica chim. Acta* **14**, 215.

Martinez-Carrion, M., Riva, F., Turano, C. and Fasella, P. (1965). Multiple forms of supernatant glutamate aspartate transaminase from pig heart. *Biophys. biochem. Res. Comm.* **20**, 206.

Martinez-Carrion, M., Turano, C., Chiancone, E., Bossa, F., Giartosio, A., Riva, F. and Fasella, P. (1967). Isolation and characterization of multiple forms of glutamate-aspartate aminotransferase from pig heart. *J. biol. Chem.* **242**, 2397.

Rogers, L. J., Shah, S. P. and Goodwin, T. W. (1966). Mevalonate-kinase isoenzymes in plant cells. *Biochem. J.* **100**, 14.

Rosalki, S. B. (1965). Creatine phosphokinase isoenzymes. *Nature, Lond.* **207**, 414.

Schapira, F. (1966). Changes in the isozymes of muscular creatine-kinase during atrophy. *C.r. Acad. Sci., Paris* **262**, 2291.

Sjovall, K. and Jergil, B. (1966). Coupled enzyme reactions in polyacrylamide gels: isoenzymes of serum ATP: creatine phosphotransferase. *Scand. J. clin. Lab. Invest.* **18**, 550.

Thoai, N. V., Thiem, N. V., Lacombe, G. and Roche, J. (1966). Heteroenzymes of adenosine-5-triphosphotic acid: L-arginine phosphotransferase. *Biochim. biophys. Acta* **122**, 547.

CHAPTER IV

Arai, M. (1966). Acid phosphatase isozymes of the synovial fluid cells in humans. *Med. biol., Tokyo* **73**, 93.

Bascur, L., Cabello, J., Veliz, M. and Gonzalez, A. (1966). Molecular forms of human-liver arginase. *Biochim. biophys. Acta* **128**, 149.

Beckman, L., Bjorling, G. and Heiken, A. (1966). Human alkaline phosphatases and the factors controlling their appearance in serum. *Acta genet. Statist. med.* **16**, 305.

Beckman, L., Bergman, S. and Lundgren, E. (1967). Isozyme variations in human cells grown in vitro. *Acta genet. Statist. med.* **17**, 304.

Behal, F. J., Klein, R. A. and Dawson, F. B. (1966). Separation and characterization of aminopeptidase and arylamidase components of human liver. *Archs Biochem. Biophys.* **115**, 545.

Berk, J. E. (1967). Serum amylase and lipase. Newer perspectives. *J. Am. med. Ass.* **199**, 98.

Bernsohn, J., Barron, K. D., Doolin, P. F., Hess, A. R. and Hedrick, M. T. (1966). Subcellular localization of rat brain esterases. *J. Histochem. Cytochem.* **14**, 455.

Bulmer, D. and Fisher, A. W. F. (1967). Inhibition of rat liver esterases by oragnophosphorous compounds. *Nature, Lond.* **213**, 202.

Corman, G. C. and Dessauer, H. C. (1967). The relationships of Anolis of the roquet species group: I. electrophoretic comparison of blood proteins. *Comp. Biochem. Physiol.* **19**, 845.

Cory, R. P. and Wold, F. (1966). Isolation and characterization of enolase from rainbow trout (Salmo Gairdnerii Gairdnerii). *Biochemistry, N.Y.* **5**, 3181.

Cory, J. G., Weinbaum, G. and Suhadolnik, R. J. (1967). Multiple forms of calf serum adenosine deaminase. *Archs Biochem. Biophys.* **118**, 428.

Dabich, D. and Neuhaus, O. W. (1966). The source of synovial fluid alkaline phosphatase. *Proc. Soc. exp. Biol. Med.* **123**, 584.

Dave, P. J., Kaplan, R. W. und Pfleiderer, G. (1966). Wirkung von Mutationen auf die Isoenzyme der Enolase aus Hefe und andere glykolytische Enzyme. *Biochem. Z.* **345**, 440.

Downey, W. K. and Andrews, P. (1966). Studies on the properties of cows milk tributyrinases and their interaction with milk proteins. *Biochem. J.* **101**, 651.

Duff, T. A. and Coleman, J. E. (1966). Macaca mulata carbonic anhydrase, crystallization and physicochemical and enzymatic properties of two isozymes. *Biochemistry, N.Y.* **5**, 2009.

Eaton, R. H. and Moss, D. W. (1967). Inhibition of the orthophosphatase and pyrophosphatase activities of human alkaline-phosphatase preparations. *Biochem. J.* **102**, 917.

Ecobichan, D. J. (1967). Hydrolases and dehydrogenases of chicken tissues. *Can. J. Biochem. Physiol.* **44**, 1277.

Ecobichan, D. J. and Israel, Y. (1967). Characterization of the esterases from electric tissue of electrophorus by starch-gel electrophoresis. *Can. J. Biochem. Physiol.* **45**, 1099.

Etzler, M. E. and Law, G. R. (1967). Effect of neuraminidase on isozymes of alkaline phosphatase and leucine aminopeptidase. *Science, N.Y.* **157**, 721.

Fenton, M. R. and Richardson, K. E. (1967). Isolation of three acid phosphatase isozymes from human erythrocytes. *Archs Biochem. Biophys.* **120**, 332.

Fishman, W. H. and Ghosh, N. K. (1967). "Isoenzymes of Human Alkaline Phosphatase". *In* "Advances in Clinical Chemistry" Vol. 10, p. 255 (O. Bodansky and C. P. Stewart, eds). Academic Press, New York.

Franzini, C. (1965). Electrophoretic behaviour of human urinary amylase. *J. clin. Path.* **18**, 664.

Fric, P. and Lojda, Z. (1966). Electrophoresis of lactate dehydrogenase, alkaline phosphatase and non-specific esterase in jejunal biopsies. *Gastroenterologia* **106**, 65.

Himoe, A., Parks, P. C. and Hess, G. P. (1967). Investigations of the chymotrypsin-catalyzed hydrolysis of specific substrates. 1. The pH dependence of the catalytic hydrolysis of N-acetyl-L-tryptophanamide by three forms of the enzyme at alkaline pH. *J. biol. Chem.* **242**, 919.

Hopkinson, D. A. and Harris, H. (1967). Column chromatography of human red cell acid phosphatase. *Ann. hum. Genet.* **31**, 29.

Inglis, N. I., Krant, M. J. and Fishman, W. H. (1967). Influence of a fat-enriched meal on human serum (L-phenylalanine sensitive) intestinal alkaline phosphatase. *Proc. Soc. exp. Biol. Med.* **124**, 699.

Ipata, P. L. (1967). Resolution of 5-nucleotidase of sheep brain from non specific phosphatase and its inhibition by nucleoside triphosphates. *Nature, Lond.* **214**, 618.

Joseph, R. R., Olivero, E. and Ressler, N. (1966). Electrophoretic study of human isoamylases. *Gastroenterology* **51**, 377.

Joshi, J. G., Hooper, J., Kuwaki, T., Sakurada, T., Swanson, J. R. and Handler, P. (1967). Phosphoglucomutase. V. Multiple forms of Phosphoglucomutase. *Proc. nat. Acad. Sci., U.S.A.* **57**, 1482.

Kleiner, H. and Graff, G. (1966). Presence of 3 isozymes hydrolyzing L-Leucyl-Beta-Napthylamide in normal human serum. *C.r. Soc. Biol., Paris* **160**, 714.

Kleiner, H. and Schramm, E. (1966). Separation des isozymes hydrolysant la l-leucyl-β-naphthylamide par electrophorese verticule en gel d'acrylamide. *Clinica chim. Acta* **14**, 377.

Leone, C. A. and Anthony, R. L. (1966). Serum esterases among registered breeds of dogs and revealed by immunoelectrophoretic comparisons. *Comp. Biochem. Physiol.* **18**, 359.

Lundin, L. G. and Allison, A. C. (1966). Acid phosphatases from different organs and animal forms compared by starch-gel electrophoresis. *Biochim. biophys. Acta* **127**, 527.

Makinen, K. K. (1967). Studies on oral enzymes. IV. Fractionation and characterisation of various hydrolytic enzymes in human saliva. *Acta odont. scand.* **24**, 709.

Moog, F. and Grey, R. D. (1967). Spatial and temporal differentiation of alkaline phosphatase on the intestinal villi of the mouse. *J. Cell Biol.* **32**, Cl.

Nüske, R. and Venner, H. (1966). Die Isoenzymfraktionen der unspezifischen alkalischen Phosphomonesterase. *Biochem. Z.* **346**, 226.

Ogawa, Y. and Matsutani, M. (1966). Study of serum leucine aminopeptidase and alkaline phosphatase isozyme by agar electrophoresis. *Jap. J. Clin. Med.* **24**, 565.

Parr, C. W. and Carter, N. D. (1967). Isoenzymes of phosphoglucose isomerase in mice. *Nature, Lond.* **216**, 511.

Penhoet, E., Rajkumar, T. and Rutter, W. J. (1966). Multiple forms of fructose diphosphate aldolase in mammalian tissues. *Proc. nat. Acad. Sci., U.S.A.* **56**, 1275.

Pfleiderer, G., Neufahrt-Kreiling, A., Kaplan, R. W. and Fortnagel, P. (1966). Biochemische und serologische Untersuchungen an multiplen Formen der Hefeenolase. *Biochem. Z.* **346**, 269.

Posen, S. (1967). Alkaline phosphatase. *Ann. intern. Med.* **67**, 183.

Posen, S., Neale, F. C., Birkett, D. J. and Brudenell-Woods, J. (1967). Intestinal alkaline phosphatase in human serum. *Am. J. clin. Path.* **48**, 81.

Rahman, Y. E. (1966). Existence of a third ribonuclease in rat liver particulates. *Biochim. Biophys. Acta* **119**, 470.

Sadihiro, R., Takanashi, S. and Kawada, M. (1965). Studies on the isozyme of β-glucuronidase. *Neurology* **58**, 104.

Scheiffarth, F., Gotz, H. and Ebert, S. (1966). Esterase-active fractions in extracts of various human organs. *Clinica chim. Acta* **14**, 519.

Schwark, W. S. and Ecobichan, D. J. (1967). Characterization of rat liver and kidney esterases. *Can. J. Biochem. Physiol.* **45**, 451.

Searcy, R. L., Hayashi, S., Berk, J. E. and Stern, H. (1966). Electrophoretic behavior of serum amylase in various mammalian species. *Proc. Soc. exp. Biol. Med.* **122**, 1291.

Secchi, T. C. and Cirla, R. (1964). Characterization of albumin esterase in a bisalbuminemic serum. *Boll. Soc. ital. Biol. sper.* **40**, 2071.

Taljedal, I. B. (1967). Electrophoretic studies on phosphatases from the pancreatis islets of obese-hyperglycaemic mice. *Acta Endocr., Copenh.* **55**, 153.

Van Ros, G. and Druet, R. (1966). Uncommon electrophoretic patterns of serum cholinesterase (pseudocholinesterase). *Nature, Lond.* **212**, 543.

Walsh, K. A., Ericsson, L. H. and Neurath, H. (1966). Bovine carboxypeptidase A variants resulting from allelomorphism. *Proc. nat. Acad. Sci., U.S.A.* **56**, 1339.

Wolf, R. O. and Taylor, L. L. (1967). Isoamylases of human saliva. *Nature, Lond.* **213**, 1128.

Yunis, A. A. and Arimura, G. K. (1966). Enzymes of glycogen metabolism in white blood cells. 3. Heterogeneity of glycogen phosphorylase from rat chloroma. *Biochim. biophys. Acta* **118**, 335.

Zech, R. and Engelhard, H. (1965). Preparative Trennung und Charakterisungen von Cholinesterasen aus Pferdeserum. *Biochem. Z.* **343**, 86.

CHAPTER V

Adams, B. and Rosso, G. (1966). Constitutive and inducible isozymes of alphaketoglutaric semialdehyde dehydrogenase in *pseudomonas*. *Biochem. biophys. Res. Commun.* **23**, 633.

K

Strehler, B. L., Hendley, D. D. and Hirsch, G. P. (1967). Evidence on a codon restriction hypothesis of cellular differentiation: multiplicity of mammalian leucyl-SRNA-specific synthetases and tissue-specific deficiency in an alanyl-SRNA synthetase. *Proc. nat. Acad. Sci., U.S.A.* **57**, 1751.

CHAPTER VI

Arends, T., Davies, D. A. and Lehmann, H. (1967). Absence of variants of usual serum pseudocholinesterase (acylcholine acylhydrolase) in South American Indians. *Acta genet. Statist. med.* **17**, 1.

Ashton, G. C. and Simpson, N. E. (1966). C5 types of serum cholinesterase in a Brazilian population. *Am. J. hum. Genet.* **18**, 438.

Beckman, L. and Wetterberg, L. (1967). Genetic and drug-induced variations in serum naphthylamidase isozymes. *Acta genet. Statist. med.* **17**, 314.

Bonsignore, A., Fornaini, G., Leoncini, G. and Fantoni, A. (1966). Electrophoretic heterogeneity of erythrocyte and leucocyte glucose-6-phosphate dehydrogenase in Italians from various ethnic groups. *Nature, Lond.* **211**, 876.

Brewer, G. J., Bowbeer, D. R. and Tashian, R. E. (1967). The electrophoretic phenotypes of red cell phosphoglucomutase, adenylate kinase, and acid phosphatase in the American negro. *Acta genet. Statist. Med.* **17**, 97.

Carter, N. D., Fildes, R. A., Fitch, L. J. and Parr, C. W. (1968). Genetically determined electrophoretic variations of human phosphogluconate dehydrogenase. *Acta genet. Statist. med.* **18**, 109.

Courtright, J. B., Imberski, R. B. and Ursprung, H. (1966). The genetic control of alcohol dehydrogenase and octanol dehydrogenases isoenzymes in Drosophila. *Genetics, Princeton* **54**, 1251.

Davidson, R. G. (1967). Electrophoretic variants of human-6-phosphogluconate dehydrogenase: population and family studies and description of a new variant. *Ann. hum. Genet.* **30**, 355.

Gordon, H., Keraan, M. M. and Vooijs, M. (1967). Variants of 6-phosphogluconate dehydrogenase within a community. *Nature, Lond.* **214**, 466.

Harris, H. (1966). Enzyme polymorphisms in man. *Proc. R. Soc. (B).* **164**, 298.

Harris, H. (1966). Genes and enzymes in man. *Cancer Res.* **26**, 2054.

Henderson, N. S. (1966). Isozymes and genetic control of NADP-malate dehydrogenase in mice. *Archs. Biochem. Biophys.* **117**, 28.

Hope, R. M. (1966). Association between serum alkaline phosphatase variants and the R-0-1 blood group system in the Australian merino. *Aust. J. biol. Sci.* **19**, 1171.

Hope, R. M. (1966). Human serum alkaline phosphatase variants and their association with the ABO blood groups in an Australian sample. *Aust. J. exp. Biol. med. Sci.* **44**, 323.

Justice, P., Ling-Yu Shih, Gordon, J., Grossman, A. and David Yi-Yung Hsia. (1966). Characterization of leukocyte glucose-6-phosphate dehydrogenase in normal and mutant human subjects. *J. Lab. clin. Med.* **68**, 552.

Karp, G. W. and Sutton, H. E. (1967). Some new phenotypes of human red cell acid phosphatase. *Am. J. hum. Genet.* **19**, 55.

Lai, L. Y. (1966). Hereditary red cell acid phosphatase types in Australian white and New Guinea native populations. *Acta genet. Statist. med.* **16**, 313.

Lai, L. Y. (1967). Polymorphism in red cell acid phosphatase of Macaca Irus. *Acta genet. Statist. med.* **17**, 104.

Lee, G. and Robinson, J. C. (1967). Agar diffusion test for serum cholinesterase typing and influence of temperature on dubucaine and fluoride numbers. *J. med. Genet.* **4**, 19.

Lewis, W. H. P. and Harris, H. (1967). Human red cell peptidases. *Nature, Lond.* **215**, 351.

Lisker, R. and Giblett, E. R. (1967). Studies on several genetic hematological traits of Mexicans. XI. Red cell acid phosphatase and phosphoglucomutase in three Indian groups. *Am. J. hum. Genet.* **19**, 174.

Locati, G., Macchi, L. and Acerboni, F. (1967). A case of complete absence of pseudocholinesterase. *Minerva med., Roma* **58**, 972.

Luffman, J. E. and Harris, H. (1967). A comparison of some properties of human red cell acid phosphatase in different phenotypes. *Ann. hum. Genet.* **30**, 387.

Modiano, G., Filippi, G., Brunelli, F., Frattaroli, W. and Siniscalco, M. (1967). Studies on red cell acid phosphatases in Sardinia and Rome. Absence of correlation with part malarial morbidity. *Acta genet. Statist. med.* **17**, 17.

Morrison, W. J. and Wright, J. E. (1966). Genetic analysis of three lactate dehydrogenase isozyme systems in trout: evidence for linkage of genes coding subunits A and B. *J. exp. Zool.* **163**, 259.

Mourant, A. E. and Tills, D. (1967). Phosphoglucomutase frequencies in Habbamite Jews and Icelanders. *Nature, Lond.* **214**, 810.

Ohno, S., Klein, J., Poole, J., Harris, C., Destree, A. and Morrison, M. (1967). Genetic control of lactate dehydrogenase formation in the hagfish Eptatretus stoutii. *Science, N.Y.* **156**, 96.

Parr, C. W. and Fitch, L. J. (1967). Inherited quantitative variations of human phosphogluconate dehydrogenase. *Ann. hum. Genet.* **30**, 339.

Rabinowitz, Y. and Dietz, A. (1967). Genetic control of lactate dehydrogenase and malate dehydrogenase isozymes in cultures of lymphocytes and granulocytes: effect of addition of phytohemagglutinin, actinomycin D or puromycin. *Biochim. biophys. Acta* **139**, 254.

Robinson, J. C. and Goldsmith, L. A. (1967). Genetically determined variants of serum alkaline phosphatase: a review. *Vox Sang.* **13**, 289.

Robson, E. B. and Harris, H. (1966). Further data on the incidence and genetics of the serum cholinesterase phenotype, C5. *Ann. hum. Genet.* **29**, 403.

Robson, E. B., Sutherland, I. and Harris, H. (1966). Evidence for linkage between the transferrin locus (TF) and the serum cholinesterase locus (E 1) in man. *Ann. hum. Genet.* **29**, 325.

Ruddle, F. H. (1966). Kidney esterases of the mouse (mus musculus): electrophoretic analysis of inbred lines C57/BL/6J, RF/J and SJL/J. *J. Histochem. Cytochem.* **14**, 25.

Shreffler, D. C. (1966). Relationship of alkaline phosphatase levels in intestinal mucosa to ABO and secretor blood groups. *Proc. Soc. exp. Biol. Med.* **123**, 423.

Stamatoyannopoulos, G., Yoshida, A., Bacopoulos, C. and Motulsky, A. G. (1967). Athens variant of glucose-6-phosphate dehydrogenase. *Science, N.Y.* **157**, 831.

Stolbach, L. L., Krant, M. J., Inglis, N. I. and Fishman, W. H. (1967). Correlation of serum L-phenylalanine-sensitive alkaline phosphatase, derived from intestine, with the ABO blood groups of cirrhotics. *Gastroenterology*, **52**, 819.

Szeinberg, A., Pipano, S., Ostfeld, E. and Eviatar, L. (1966). The silent gene for serum pseudocholinesterase. *J. med. Genet.* **3**, 190.

Thuline, H. C., Morrow, A. C., Norby, D. E. and Motulsky, A. G. (1967). Autosomal phosphogluconic dehydrogenase polymorphism in the cat (Felis Catus L.). *Science, N.Y.* **157**, 431.

K*

Weiss, M. C. and Ephrussi, B. (1966). Studies of interspecific (rat times mouse) somatic hybrids, 11. Lactate dehydrogenase and beta-glucuronidase. *Genetics, Princeton* **54**, 1111.

Whittaker, M. (1967). The pseudocholinesterase variants. A study of fourteen families selected via the fluoride resistant phenotype. *Acta genet. Statist. med.* **17**, 1.

Yoshida, A. (1967). A single amino acid substitution (asparagine to aspartic acid) between normal (B+) and the common negro variant (A+) of human glucose-6-phosphate dehydrogenase. *Proc. natn. Acad. Sci., U.S.A.* **57**, 835.

Yoshida, A., Stamatoyannopoulos, G. and Motulsky, A. G. (1967). Negro variant of glucose-6-phosphate dehydrogenase deficiency (A−) in man. *Science, N.Y.* **155**, 97.

CHAPTER VII

Biserte, G., Farriaux, J. P., Hoste, A., Guinmard, M. P. and Fontaine, G. (1966). La lacticodeshydrogenase serique (LDH) et ses isoenzymes: technique d'etude et resultants chez l'enfant. *Annls Biol. clin.* **24**, 663.

Bush, F. M. (1967). Developmental and populational variation in electrophoretic properties of dehydrogenases, hydrolases and other blood proteins of the house sparrow, *passer domesticus*. *Comp. Biochem. Physiol.* **22**, 273.

Fisher, J. R., Curtis, J. L. and Woodward, W. D. (1967). Developmental changes and control of xanthine dehydrogenases in developing chicks. *Devl. Biol.* **15**, 289.

Genis-Galves, J. M. and Maisel, H. (1967). Lactate dehydrogenase isozymes in changes during lens differentiation in the chick. *Nature, Lond.* **213**, 283.

Hinks, M. and Masters, C. J. (1966). The ontogenetic variformity of lactate dehydrogenase in feline and cavian tissues. *Biochim. biophys. Acta* **130**, 458.

Lagnado, J. R. and Madeleine Hardy. (1967). Brain esterases during development. *Nature, Lond.* **214**, 1207.

Lowenthal, A., Karcher, D., Van Sande, M. and Wintgens, M. (1966). Distribution of the isoenzymes of lactate dehydrogenase in the brain of the human fetus and in the cerebrospinal fluid of the newborn infant. *Acta neurol. psychiat. belg.* **66**, 553.

Manwell, C., Baker, C. M. and Betz, T. W. (1966). Ontogeny of haemoglobin in the chicken. *J. Embryol. exp. Morph.* **16**, 65.

McWhinnie, D. J. and Saunders, J. W. Jr. (1966). Developmental patterns and specificities of alkaline phosphatase in the embryonic chick limb. *Devl. Biol.* **14**, 169.

Patton, G. W., Mets, L. and Villee, C. A. (1967). Malic dehydrogenase isozymes: distribution in developing nucleate and anucleate halves of sea urchin eggs. *Science, N.Y.* **156**, 400.

Pelichova, H., Koldovsky, O., Uher, J., Kraml, J., Heringova, A. and Jirsova, V. (1966). Fetal development of nonspecific esterases and alkaline phosphatases activities in the small intestine of the man. *Biologia Neonat.* **10**, 281.

Salthe, S. N. and Kaplan, N. O. (1966). Immunology and rates of enzyme evolution in the amphibia in relation to the origins of certain taxa. *Evolution* **20**, 603.

Schmukler, M. and Barrows, C. H. Jr. (1967). Effect of age on dehydrogenase heterogeneity in the rat. *J. Geront.* **22**, 8.

Ueda, Y. (1966). On lactate dehydrogenase and isozyme in the developing human embryo. *Acta paediat. jap.* **70**, 907.

Weatherall, D. J. and McIntyre, P. A. (1967). Developmental and acquired variations in erythrocyte carbonic anhydrase isoenzymes. *Br. J. Haemat.* **13**, 106.

Wright, D. A. and Moyer, F. H. (1966). Parental influences on lactate dehydrogenase in the early development of hybrid frogs in the Genus Rana. *J. exp. Zool.* **163**, 215.

CHAPTER VIII

Abramson, C. (1967). Staphylococcal hyaluronate lyase: multiple electrophoretic and chromatographic forms. *Archs Biochem. Biophys.* **121**, 103.

Ackrell, B. A. C., Asato, R. N. and Mower, E. F. (1966). Multiple forms of bacterial hydrogenases. *J. Bacteriol.* **92**, 828.

Andreev, L. N. and Shaw, M. (1965). A note on the effect of rust infection on peroxidase isozymes in flax. *Can. J. Bot.* **43**, 1479.

Burns, J. M., and Johnson, F. M. (1967). Esterase polymorphism in natural populations of a sulfur butterfly, Colias eurytheme. *Science, N.Y.* **156**, 93.

Chesbro, W. R., Stuart, D. and Burke, J. J. (1966). Multiple molecular weight forms of staphylococcal nuclease. *Biochem. Biophys. Res. Commun.* **23**, 783.

Desborough, S. and Peloquin, S. J. (1967). Esterase isozymes from solanum tubers. *Phytochemistry* **6**, 989.

Doane, W. W. (1967). Quantitation of amylases in *Drosophila* separated by acrylamide gel electrophoresis. *J. exp. Zool.* **164**, 363.

Dunagan, T. T. and de Luque, O. (1967). Isozyme patterns for lactic and malic dehydrogenases in Macracanthorhynchus hirudinaceus. *J. Parasitol.* **52**, 727.

Eguchi, M. and Narumi, Y. (1966). Genetics studies on isozymes of the integument esterase in the silkworm Bombyx mori. *Jap. J. Genet.* **41**, 267.

Eguchi, M. and Yoshitake, N. (1967). Electrophoretic variation of proteinase in the digestive juice of the silkworm *Bombyx mori L. Nature, Lond.* **214**, 843.

Endo, T. and Schwartz, D. (1966). Tissue specific variations in the urea sensitivity of the E_1 esterase in maize. *Genetics, Princeton* **54**, 233.

Fleming, L. W. and Duerkson, J. D. (1967). Evidence for multiple molecular forms of yeast β-galactosidase in a hybrid yeast. *J. Bacteriol.* **93**, 142.

Fottrell, P. F. (1966). Dehydrogenase isoenzymes from legume root nodules. *Nature, Lond.* **210**, 198.

Fredrick, J. F. (1964). Polyacrylamide gel studies of the isozymes involved in polyglucoside synthesis in the algae. *Phyton. Reg. Int. Bot. Exp.* **21**, 85.

Frederick, J. F. (1967). Glucosyltransferase isozymes in algae. *Phytochemistry* **6**, 1041.

Giacomelli. M. (1967). Peroxidase isozymes in barley leaves from irradiated seeds. *Nature, Lond.* **213**, 1265.

Grimes, H. and Fottrell, P. F. (1966). Enzymes involved in glutamate metabolism in legume root nodules. *Nature, Lond.* **212**, 295.

Hall, R. (1967). Proteins and catalase isoenzymes from *fusarium solani* and their taxonomic significance. *Aust. J. biol. Sci.* **20**, 419.

Johnson, F. M., Kanapi, C. G., Richardson, R. H., Wheeler, M. R. and Stone, W. S. (1966). An analysis of polymorphisms among isozyme loci in dark and light drosophila ananassae strains from American and Western Samoa. *Proc. natn Acad. Sci.* **56**, 119.

Johnson, F. M., Kanapi, C. G., Richardson, R. H. and Sakai, R. K. (1967). Isozyme variability in species of the genus *Drosophila*. I. A multiple allelic isozyme system in *Drosophila busckii*: inheritance and general considerations. *Biochemical Genetics* **1**, 35.

Kay, E., Shannon, L. M. and Lew, J. Y. (1967). Peroxidase isozymes from horseradish roots. 11. Catalytic properties. *J. biol. Chem.* **242**, 2470.

Kellen, J. (1965). Isoenzymes of lactate dehydrogenase in micro-organisms. *Nature, Lond.* **207**, 783.

Hubby, J. L. and Lewontin, R. C. (1966). A molecular approach to the study of genic heterozygosity in natural populations. 1. The number of alleles at different loci in *Drosophila pseudoobscura*. *Genetics, Princeton* **54**, 577.

Knowles, B. B. and Fristrom, J. W. (1967). The electrophoretic behaviour of ten enzyme systems in the larval integument of *drosophila melanogaster*. *J. Insect. Physiol.* **13**, 731.

Kojima, Ken-ichi and Yarbrough, K. M. (1967). Frequency-dependent selection at the esterase 6 locus in *Drosophila melanogaster*. *Proc. natn. Acad. Sci., U.S.A.* **57**, 645.

Lewontin, R. C. and Hubby, J. L. (1966). A molecular approach to the study of genic heterozygoisty in natural populations. 11. Amount of variation and degree of heterozygosity in natural populations of *Drosophila pseudoobscura*. *Genetics, Princeton*, **54**, 595.

Marshall, K. C. (1967). Electrophoretic properties of fast and slow growing species of rhizobium. *Aust. J. biol. Sci.* **20**, 429.

McReynolds, M. S. (1967). Homologous esterases in three species of the virilis group of Drosophila. *Genetics, Princeton* **56**, 527.

MacIntyre, R. J. and Dean, M. R. (1967). Sub-units of acid phosphatase-1 in *Drosophila melanogaster*: reversible dissociation in vitro. *Nature, Lond.* **214**, 274.

Ockerse, R., Ziegel, B. Z. and Galston, A. W. (1966). Hormone induced repression of peroxidase isozyme in plant tissue. *Science, N.Y.* **151**, 452.

Pandey, K. K. (1967). Origin of genetic variability: combinations of peroxidase isozymes determine multiple allelism of the S gene. *Nature, Lond.* **213**, 669.

Racusen, D. and Foote, M. (1966). Peroxidase isozymes in bean leaves by preparative disc electrophoresis. *Can. J. Bot.* **44**, 1633.

Scandalios, J. G. (1966). Amylase isozyme polymorphism in maize. *Planta* **69**, 244.

Schlesinger, M. J. (1967). The reversible dissociation of the alkaline phosphatase of escherichia coli. 3. Properties of antibodies directed against the subunit. *J. biol. Chem.* **242**, 1599.

Schlesinger, M. J. (1967). Formation of a defective alkaline phosphatase subunit by a mutant of escherichia coli. *J. biol. Chem.* **242**, 1604.

Schwartz, D. (1967). E_1 esterase isozymes of maize: on the nature of the gene-controlled variation. *Proc. natn. Acad. Sci., U.S.A.* **58**, 568.

Shannon, L. M., Kay, E. and Lew, J. Y. (1966). Peroxidase isozymes from horse-radish roots. I. Isolation and physical properties. *J. biol. Chem.* **241**, 2166.

Sherman, I. W. (1966). Malic dehydrogenase heterogeneity in malaria (Plasmodium lophurae, and P. Berghei). *J. Protozool.* **13**, 344.

Siegel, B. Z. and Galston, A. W. (1967). The isoperoxidases of Pisum sativum. *Pl. Physiol.* **42**, 221.

Solymosy, F., Szirmai, J., Beczner, L. and Farkas, G. L. (1967). Changes in peroxidase-isozyme patterns induced by virus infection. *Virology* **32**, 117.

Ting, I. P., Sherman, I. W. and Dugger, W. M. Jr. (1966). Intracellular location and possible function of malic dehydrogenase isozymes from young maize root tissue. *Pl. Physiol.* **41**, 1083.

Trevithick, J. R. and Metznberg, R. L. (1964). The invertase isozyme formed by neurospora protoplasts. *Biochem. Biophys. Res. Commun.* **16**, 319.

Trevithick, J. R. and Metzenberg, R. L. (1966). Molecular sieving by neurospora cell walls during secretion of invertase isozymes. *J. Bact.* **92**, 1010.

Wallace, B. J. and Pittard, J. (1967). Genetic and biochemical analysis of the iso-enzymes concerned in the first reaction of aromatic biosynthesis in E. Coli. *J. Bact.* **93**. 237.

Weimberg, R. (1967). Effect of sodium chloride on the activity of a soluble malate dehydrogenase from pea seeds. *J. biol. Chem.* **242**, 3000.

Wright, C. A., File, S. K. and Ross, G. C. (1966). Studies on the enzyme systems of planorbid snails. *Ann. trop. Med. Parasit.* **60**, 522.

Yoshitake, N. (1966). Difference in the multiple forms of several enzymes between wild and domestic silkworms. *Biol. Abs.* **48**, 20548.

Yu, CH-T and Rappaport, H. P. (1966). Multiple forms of leucyl-transfei ribonucleic acid synthetase activity from Escherichia coli. *Biochim. biophys. Acta* **123**, 134.

Yue, S. B. (1966). Isoenzymes of malate dehydrogenase from barley seedlings. *Phytochem.* **5**, 1147.

CHAPTER IX

Aber, C. P., Noble, R. L., Thompson, G. S. and Wyn-Jones, E. (1966). Serum lactic dehydrogenase isoenzymes in myxoedema heait disease. *Br. Heart J.* **28**, 663.

Aber, C. P., Brunt, P. E., Jones, E. W., Richards, T. G. and Short, A. H.(1966). Liver function after myocardial infarction. *Lancet* **i**, 1391.

Batsakis, J. G. and Siders, D. (1967). Enzyme molecular heterogeneity (isoenzymes) in surgical diagnosis. *Archs Surg., Chicago* **95**, 138.

Bohn, L., Bendixen, G. and Hjermind, P. (1966). Haemning af serum-laktatdehdrogenase-aktiviteten med urinstof ved diagnosen af hjerteinfarkt. *Nord. Med.* **76**, 1467.

Cao, A., Macciotta, A., Fiorelli, G., Mannucci, P. M., and Ideo, G. (1966). Chromatographic and electrophoretic pattern of lactate and malate dehydrogenase in normal human adult and foetal muscle and in muscle of patients affected by Duchenne muscular dystrophy. *Enzymol. biol. clin.* **7**, 156.

Fric, P. and Lojda, Z. (1966). Electrophoresis of lactate dehydrogenase, alkaline phosphatase and non-specific esterase in jejunal biopsies of controls and patients with malabsorption syndrome. *Gastroenterologia* **106**, 65.

Fric, P., Lojda, Z. and Pokorna, M. (1966). Lactate dehydrogenase isozymes in jejunal biopsies of children with celiac disease. *J. Pediat.* **68**, 813.

Fujimoto, M. (1965). Symposium on the significance of enzymes in the field of internal medicine. 3. Amylase and lipase activities in pancreatic impairment. *J. Japan Soc. Intern. Med.* **54**, 718.

Ganrot, P. O. (1967). Lupoid cirrhosis with serum lactic acid dehydrogenase linked to an α immunoglobulin. *Experientia* **23**, 593.

Gerhardt, W., Clausen, J., Christensen, E. and Riishede, J. (1967). Lactate dehydrogenase isoenzymes in the diagnosis of human benign and malignant brain tumours. *J. natn. Cancer Inst.* **38**, 343.

Godwin, E. L., Rees, K. R. and Varcoe, J. S. (1967). Nuclear RNA synthesis in rat liver during the early stages of chemical carcinogenesis. *Br. J. Cancer* **21**, 166.

Goldberg, A. F., Takakura, K., and Rosenthal, R. (1966). Electrophoretic separation of serum acid phosphatase isoenzymes in Gaucher's disease, prostatic carcinoma and multiple myeloma. *Nature, Lond.* **211**, 41.

Hellman, B. and Taljedal, I. S. (1967). Quantitative studies on isolated pancreatic islets of mammals, activity and heterogeneity of lactate dehydrogenase in obese-hypoglycemic mice. *Endocrinology* **81**, 124.

Hershey, F. B., Johnston, G., Murphy, S. M. and Schmitt, M. (1966). Pyridine nucleotide linked dehydrogenases and isozymes of normal rat breast and growing and regressing breast cancers. *Cancer Res.* **26**, 265.

Huser, H. J. (1967). Diagnostic significance of isoenzymes in hematologic diseases. *Dt. med. Wschr.* **92**, 1313.

Igarashi, M., Mikami, R. and Nagano, M. (1966). Blood alkaline phosphatase isozyme and osteoarticular diseases. *Orthop. Surg., Tokyo* **17**, 245.

Ikawa, Y. (1965). Isozyme histochemistry, especially in relation to the malignant transformation and differentiation: histochemical demonstration of lactic dehydrogenase isozymes. *Gann. Jap. J. Cancer Res.* **56**, 201.

Inglis, N. I., Krant, M. J. and Fishman, W. H. (1967). Influence of a fat-enriched meal on human serum (L-Phenylalanine sensitive) intestinal alkaline phosphatase. *Proc. Soc. exp. Biol. Med.* **124**, 699.

Jacey, M. J. and Schaefer, K. E. (1967). Regulation of plasma lactic dehydrogenases in chronic respiratory acidosis. *Am. J. Physiol.* **212**, 859.

Kar, N. C. and Pearson, C. M. (1967). Electrophoretic patterns of several dehydrogenases and hydrolases in muscles in human myopathies. *Am. J. clin. Path.* **47**, 594. *Proc. Soc. exp. Biol. Med.* **124**, 699.

Katsunuma, N. (1965). Symposium on the significance of enzymes in the field of internal medicine. Clinical significance of transaminase isozyme. *J. Japan Soc. Intern. Med.* **54**, 750.

Kontinnen, A. and Lindy, S. (1965). Denaturation of lactic dehydrogenase isozymes and its clinical application. *Nature, Lond.* **208**, 782.

Kontinnen, A. and Lindy, S. (1967). Assay of cardiac lactate dehydrogenase isoenzymes by means of urea. *Acta med. scand.* **161**, 513.

Lewandowicz, J. and Kedziora, J. (1966). Behavior of isoenzymes of lactic dehydrogenase in rheumatic heart disease. *Polskie Archwm. Med. wewn.* **37**, 283.

Lindy, S. and Kontinnen, A. (1967). Urea-stable lactate dehydrogenase as an index of cardiac isoenzymes. *Am. J. Cardiol.* **19**, 563.

Lindy, S. and Kontinnen, A. (1967). Clinical applicability of urea-LDH test in various pyruvate concentration. *Clinica chim. Acta* **17**, 223.

Messer, R. H. (1967). Heat-stable alkaline phosphatase as an index of placental function. *Am. J. Obstet. Gynec.* **98**, 659.

Miyoshi, K. (1965). Symposium on the significance of enzymes in the field of internal medicine. Abnormality of myoglobin subfractions and creatinine kinase isozyme pattern in skeletal muscles of Duchenne type of progressing muscular dystrophy. *J. Japan Soc. Intern. Med.* **54**, 759.

Nakagawa, S. and Tsuji, H. (1966). Multiplicity of leucinamide-splitting enzymes in normal and hepatitis sera. *Clinica chim. Acta* **13**, 155.

Newton, M. A. (1967). The clinical application of alkaline phosphatase electrophoresis. *Q.Jl Med.* **36**, 17.

Oda, T. (1965). Symposium on the significance of enzymes in the field of internal medicine. 1. On the origin of serum enzyme i.e. phosphomonoesterase and isocitrate dehydrogenase, with special reference to those isodynamic enzyme or isozyme. *J. Japan Soc. Intern. Med.* **54**, 712.

Ogawa, Y. and Matsutani, M. (1966). Study of serum leucine aminopeptidase and alkaline phosphatase isozyme by agar electrophoresis and its diagnostic significance. *Jap. J. Clin. Med.* **24**, 565.

Oldershausen, H. F., Von Luttje, A., Eggstein, M. and Zysno, E. (1965). On the serum enzyme diagnosis of muscular disorders in intoxications and infectious diseases. *Verh. dt. Ges. inn. Med.* **71**, 653.

Pearson, C. M. and Kar, N. C. (1966). Isoenzymes: general considerations and alterations in human and animal myopathies. *Ann. N.Y. Acad. Sci.* **138**, 293.

Rabinowitz, Y. and Dietz, A. A. (1967). Malic and lactic dehydrogenase isozymes of normal and leukemic leukocytes separated on glass bead columns. *Blood* **29**, 182.

Reeves, A. L., Busby, E. K. and Scotti, L. (1966). Gel electrophoresis in the study of pneumoconioses. *Am. ind. Hyg. Ass. J.* **27**, 278.

Ringoir, S., Derom, F, Schoofs, E. and Van Acker, K. (1967). Niertransplantatie met lijkennieren: 5 gevallen. *Nederlands Tijdschrift voor Geneeskune, Jaargang* 111, Nr. 1.

Schapira, F. (1966). Aldolase isozymes in cancer. *Europ. J. Cancer* **2**, 131.

Schapira, F., Dreyfus, J. C. and Schapira, G. (1966). Fetal-like patterns of lactic dehydrogenase and aldolase isozymes in some pathological conditions. *Enzymol. biol. clin.* **7**, 98.

Scheider, K. W., Lehmann, F. G. and Schering, G. (1966). Die enzymatische Diagnostik des Herzinfarktes III LDH isoenzyme beim Herzinfarkt. *Enzymol biol. clin.* **6**, 52.

Spector, I., McFarland, W., Trujillo, N. P. and Ticktin, H. E. (1966). Bone marrow lactic dehydrogenase in haematologic and neoplastic disease. *Enzymol. biol. clin.* **7**, 78.

Stolbach, L. L., Krant, M. J., Inglis, N. I. and Fishman, W. H. (1967). Correlation of serum L-phenylalanine-sensitive alkaline phosphatase, derived from intestine, with the ABO blood group of cirrhotics. *Gastroenterology* **52**, 819.

Streicher, E., Jachnecke, J., Loffler, G. and Meisch, M. (1965). On the differential diagnosis of acute anuria. *Verh. dt. Ges. inn. Med.* **71**, 713.

Takahashi, H. (1966). Clinical significance of serum alkaline phosphatase, leucine aminopeptidase and LDH isozyme. *Jap. J. Clin. Path.* **14**, 417.

Trujillo, N. P., Nutter, D. and Evans, J. M. (1967). The isoenzymes of lactic dehydrogenase. II. Pulmonary embolism, liver disease, the postoperative state, and other medical conditions. *Archs intern. Med.* **119**, 333.

Tsubura, E., Aoki, R., Masaki, S., Ogawa, M. and Tatsuta, M. (1966). Diagnosis of cancer and isozymes. 3. Trial of enzymic diagnosis of cancer by beta-glucuronidase isozyme. *Jap. J. clin. Med.* **24**, 377.

Yamamura, Y., Tsubura, E., Masaki, S., Ogawa, M. and Tatsuta, M. (1966). Biochemical review of lung cancer. *Jap. J. clin. Med.* **24**, 421.

CHAPTER X

Homolka, J., Kristanova, D. and Vecerek, B. (1966). The use of nitro blue tetrazolium for the visualisation of isozymes of lactic dehydrogenase. *Clinica chim. Acta* **13**, 125.

Lance, B. M. (1966). A method for separation and quantitation of lactate dehydrogenase isoenzymes on cellulose acetate. *Am. J. clin. Path.* **46**, 401.

Pietruszko, R. and Baron, D. N. (1967). A staining procedure for demonstration of multiple forms of aldolase. *Biochim. biophys. Acta* **132**, 203.

Rattazzi, M. C., Bernini, L. F., Fiorelli, G. and Mannucci, P. M. (1967). Electrophoresis of glucose-6-phosphate dehydrogenase: a new technique. *Nature, Lond.* **213**, 79.

Warburton, F. G. and Waddecar, J. (1966). Use of cellulose acetate gel (Cellogel) for the determination of lactic dehydrogenase isoenzymes. *J. clin. Path.* **19**, 517.

Wieme, R. J. (1966). An improved agar support for electrophoresis of LDH isoenzymes. *Clinica chim. Acta* **13**, 138.

Author Index

Numbers in italics indicate the page on which the references are listed.

A

Aalund, O., 62, *203*
Abdul-Fadl, M. A. M., 64, 65, *203*
Abelmann, W. H., 148, *214*
Aber, C. P., *237*
Abramson, C., *235*
Acerboni, F., *233*
Acheson, J., 157, *203*
Ackerman, D. R., *227*
Ackrell, B. A. C., *235*
Adam, A., 99, 100, 101, *217*
Adams, B., *231*
Adams, E., 124, 125, *203*
Aebi, H., 5, 6, 8, 47, 53, 54, 55, 83, 125, 126, 164, 183, 195, *205, 207, 216, 218, 221*
Agnall, I. P. S., 5, *203*
Agostini, A., 10, 164, 165, *203, 206, 225*
Alberty, R. A., 140, *209*
Aldridge, W. N., 66, *203*
Allen, J. M., 5, 63, 64, 83, 89, 141, 175, 180, 190, 196, 197, *203*
Allen, S. L., 141, 142, *203*
Allen, T. M., *227*
Allen, G. A., 71, *203*
Allison, A. C., 65, *214, 230*
Allison, M. J., 85, 191, *203*
Allison, W. S., 23, 24, 27, 126, 127, *222*
Allott, E. N., 109, *203*
Alpers, D. H., 145, *203*
Altland, P. D., 9, *208*
Alves, J. J. A., 68, *206*
Amelung, D. I., 148, *203*
Amore, G., 12, 13, 119, *208, 225*
Anagnostou-Kakaras, E., 57, 159, 196, *215*
Anderson, B. M., 28, *203*
Anderson, H., 164, *208*
Anderson, S., *225*
Andreev, L. N., *235*
Andrews, P., 73, 184, *203, 207, 230*
Angel, R. W., 67, 71, 199, *210*

Angeletti, P. U., 168, 169, *203, 215*
Anstall, H. B., 79, 102, 177, *203, 220*
Anthony, R. L., *230*
Aoki, R., *239*
Appel, S. H., 145, *203*
Appella, E., 17, 18, 23, 38, *203, 214*
Arai, M., *229*
Arends, T., *232*
Arfors, K. E., 60, 113, *203*
Arfshapour, F., 134, *203*
Arimura, G. K., *231*
Armstrong, A. R., 59, *203*
Arnheim, N., Jr., *225*
Asato, R. N., *233*
Ashton, G. C., *232*
Atanasov, N., *225*
Augustinsson, K-B., 66, 70, 71, 181, *203*
Avellone, S., *225*
Avrameas, S., 23, 200, *203, 205, 217*
Aw, S. E., 77, *203*

B

Bach, M. L., 59, 143, *203*
Bachhawat, B. K., 24, 58, *208, 218*
Bacopoulos, C., *233*
Baker, C. M., *234*
Baker, R. W. R., 56, 75, *204*
Balchum, D. J., *225*
Balek, R. W., *225*
Bamford, K. F., 60, 61, 109, 113, *204*
Banks, J., 99, 193, *215*
Bandoin, J., 65, *218*
Banting, F. G., 59, *203*
Bar, U., 164, 168, 169, *204*
Barka, T., 66, 197, *204*
Barnett, H., 164, 181, *204*
Barnstein, P. L., 88, *220*
Baron, D. N., 44, 45, 175, 192, *204, 239*
Barrett, K. J., 144, *207*
Barron, K. D., 66, 67, 69, 71, 163, 199, *204, 210, 229*
Barrows, C. H., Jr., *234*

241

G

Gabriel, O., 200, *217*
Gafni, D., 100, *217*
Gahne, B., 108, 115, *208*
Gajos, E., 71, 108, *211*
Gallitelli, L., 149, *219*
Galston, A. W., 138, *236*
Ganrot, P. O., *237*
Gantner, G., 50, *212*, *228*
Garber, E. D., 141, *215*
Garbus, J., 9, *208*
Gardner, P. S., 165, *213*
Gartler, S. M., 101, 102, 167, *214*
Geiger, H. K., 131, *208*
Gelboin, H. V., 166, *208*
Gelderman, A. H., 37, 166, *208*
Gelotte, B., *208*
Genest, K., 109, *211*
Genis-Galves, J. M., *234*
Gerhardt, W., 13, 164, 167, *206*, *208*
  237
German, J. L., 16, *208*
Gerszten, E., 84, 191, *203*
Gervasini, N., 150, *219*
Ghazanfar, A. S., 55, *218*
Giacomelli, M., *235*
Giartosio, A., *229*
Gibbons, B. H., 55, *218*
Giblett, E. R., 115, 116, *208*, *233*
Gibson, A., 164, *204*
Gibson, C. W., 42, 164, *223*
Gibson, D. M., 24, *208*
Giersberg, H., 4, 10, *209*
Gilbert, L. I., 134, *208*
Gilman, A., 109, *208*
Glassman, E., 129, 130, *208*, *212*, *223*
Glen-Bott, A. M., 92, 94, *206*
Glenner, G. G., 74, 75, *210*
Godwin, E. L., *237*
Goebelsman, V., 74, *208*
Goldberg, A. F., 161, *208*, *237*
Goldberg, E., 5, 10, 14, 19, 42, 95, 96,
  124, *209*
Goldberger, R. F., 17, *207*
Goldenberg, G. J., 16, *223*
Goldman, R. D., 164, *209*
Goldsmith, L. A., *233*
Goldstein, D. P., 168, *218*
Goll, R., 180, *218*

Gomes, J. M., 68, *206*
Gomori, G., 197, *209*
Gonder, M. J., 162, *219*
Gonzalez, A., *229*
Gonzalez, C., 52, *209*
Goodfriend, T., 16, 84, 85, *209*
Goodfriend, T. L., 82, 83, 84, 85, 86, 151,
  206, *209*
Goodman, M., 94, *220*
Goodwin, T. W., *229*
Gordin, R., 156, *217*
Gordon, A. H., 170, *209*
Gordon, H., *226*, *232*
Gordon, J., *232*
Gotts, R., 191, *209*
Gotz, H., *231*
Grabar, P., 23, 68, 170, 180, *209*, *215*,
  217
Graff, G., *230*
Gray, P. W. S., 109, *207*
Graymore, C. N., 13, *209*
Green, R. A., 166, 191, *220*
Green, S., 60, *208*
Greene, S. F., 57, 159, *205*
Greenslade, K., 89, *214*
Gregory, K., 1, 5, 6, 8, 37, 38, 146, 149,
  154, 175, 176, 184, 189, *223*
Gregory, K. F., 10, 18, 31, 38, 80, 165,
  185, 186, *217*, *226*, *227*
Greiling, H., 144, 151, *209*
Grell, E. H., 128, *209*
Grey, R. D., 58, *215*, *231*
Grimes, H., 137, *209*, *235*
Grimm, F. C., 43, *209*
Grivea, M., 61, 62, *204*
Grossbard, L., 53, *209*, *228*
Grossberg, A. L., 58, *209*
Grossman, A., *232*
Grossman, L. I., 1, 43, 176, 192, *220*
Grunder, A. A., 107, *209*
Grundig, E., 161, *209*
Guarneri, R., 11, 12, 126, 189, *205*
Guinmard, M. P., *234*
Gunn, D. R., 110, *211*
Gustavsson, I., 102, *216*
Gutman, A. B., 59, *209*
Gutmann, H. R., 48, *212*
Gutter, F. J., 182, *219*
Güttler, F., 152, *209*

AUTHOR INDEX

Subject Index

A

Abdominal emergencies
  serum LDH isoenzymes in, 153
  serum arylamidase isoenzymes in, 162
Acetabularia (and *Acicularia*)
  acid phosphatases, 142–143
Acetolactate metabolism
  isoenzymes and, 89
Acetone
  effect on serum LDH isoenzymes, 155, 185
Acetylcholinesterase (*see also* cholinesterase)
  serum, 71
Acid phosphatase isoenzymes
  algae, 142–143
  differential inhibitors, 64–65
  Drosophila, 132
  feline cord, 66
  genetic variation, 115–116
  hamster liver, 66
  human, adrenal, 64
    ,, brain, 66
    ,, erythrocytes, 64–65, 115–116
    ,, intestine, 64
    ,, kidney, 65–66
    ,, liver, 64–66
    ,, pancreas, 64
    ,, prostate, 64–65
    ,, seminal plasma, 65
    ,, serum, 64–65, 161–162
    ,, spleen, 64–65
    ,, thyroid, 64
  mouse liver and kidney, 66
  Phaseolus vulgaris, 138
  Protozoa, 141–142
  rabbit liver, 66
  rat liver and kidney, 66
  serum pattern in disease, 161–162
  techniques for demonstration, 197–198
Acrylamide gel (*see* electrophoresis)
Acute pancreatitis
  serum LDH isoenzymes in, 153
  serum arylamidase isoenzymes in, 162
Adenosine triphosphatase

rat liver, 64
  calf intestinal, 64
  mitochondrial, 64
Adenovirus 12,
  effect on LDH isoenzyme patterns of cultured cells, 165
Adenylate kinase
  genetic variation in human tissues, 103–104
Adrenal
  human, acid phosphatase, 64
    ,, esterases, 69
    ,, LDH isoenzyme pattern, 5–6
Aerobacter aerogenes
  malate dehydrogenases, 43
Agar (*see* electrophoresis, immuno-electrophoresis and immunological techniques)
Alanine aminotransferase
  rat liver, 88
  effect of hormones, 88
  effect of X-irradiation, 88
Alcohol
  inhibition of acid phosphatases, 64
  inhibition of alkaline phosphatases, 160, 185
Alcohol dehydrogenase isoenzymes, 47, 48, 128–129, 136
  Drosophila, 128–129
  horse liver, 47
  human liver, 47
  maize, 136
  technique for demonstration, 47, 193–194
  yeast, 47
Aldolases
  bovine liver, 79
  human tissues, 79
  immunochemical studies, 79
  kinetic characteristics, 79
  rabbit muscle, 79
  rat tissues, 79
  separation by chromatography, 79
  separation by starch gel electrophoresis, 79